Report of Investigations 9679

Recommendations for a New Rock Dusting Standard to Prevent Coal Dust Explosions in Intake Airways

Kenneth L. Cashdollar, Michael J. Sapko, Eric S. Weiss, Marcia L. Harris, Chi-Keung Man, Samuel P. Harteis, and Gregory M. Green

DEPARTMENT OF HEALTH AND HUMAN SERVICES
Centers for Disease Control and Prevention
National Institute for Occupational Safety and Health
Office of Mine Safety and Health Research
Pittsburgh, PA

May 2010

This document is in the public domain and may be freely copied or reprinted.

Disclaimer

Mention of any company or product does not constitute endorsement by the National Institute for Occupational Safety and Health (NIOSH). In addition, citations to Web sites external to NIOSH do not constitute NIOSH endorsement of the sponsoring organizations or their programs or products. Furthermore, NIOSH is not responsible for the content of these Web sites.

Ordering Information

To receive documents or other information about occupational safety and health topics, contact NIOSH at

> Telephone: **1–800–CDC–INFO** (1–800–232–4636)
> TTY: 1–888–232–6348
> e-mail: cdcinfo@cdc.gov
>
> or visit the NIOSH Web site at **www.cdc.gov/niosh**.

For a monthly update on news at NIOSH, subscribe to NIOSH *eNews* by visiting **www.cdc.gov/niosh/eNews**.

DHHS (NIOSH) Publication No. 2010–151

May 2010

SAFER • HEALTHIER • PEOPLE™

Contents

Executive Summary 1
Introduction 3
Comparison of International Rock Dusting Requirements 5
Experimental Procedures 8
    Size Data for Intake Airways 15
    Size Data for Return Airways 18
    MSHA Dust Survey Results from Intake and Return Airways 18
    Limestone Rock Dust Inerting 19
Summary 21
Recommendation 22
Acknowledgments 22
References 23
Appendix A: Analyses of Size of Coal Dust Particles from Mine Intake Airways 27
Appendix B: Analyses of Size of Coal Dust Particles from Mine Return Airways 39
Appendix C: Discussion of the Coal Dust and Rock Dust Properties and Experiments 43
    Limestone Rock Dust Inerting Discussion 44
    Effect of Particle Size on Coal Dust Explosibility 49

Figures

Figure 1. Effect of particle size of coal dust on the explosibility of Pittsburgh seam bituminous coal as tested within BEM. 5

Figure 2. MSHA Coal Mine Safety and Health Districts, identified by number. 8

Figure 3. Original analyses of coal sieve size and analyses of sieve size of acid-leached mixture containing 30% medium-sized Pittsburgh coal and 70% limestone rock dust. 10

Figure 4. Original analyses of coal sieve size and analyses of sieve size of acid-leached mixture containing 30% medium-sized Pittsburgh coal, 60% limestone rock dust, and 10% kaolin clay. 10

Figure 5. Plan view of the Lake Lynn Experimental Mine (LLEM). 12

Figure 6. Side view of LLEM A-drift and D-drift test zones for determining rock dust inerting requirements. 13

Figure 7. Placing coal and rock dust mixture on shelves in the LLEM. 14

Figure 8. Distributing test dust mixture at the LLEM. 14

Figure 9. Coal particle size by MSHA district. 16

Figure 10. Coal particle size by coal seam. 18

Figure 11. Effect of particle size of coal dust on the explosibility of Pittsburgh seam bituminous coal as tested within LLEM. 19

Figure C-1. Effect of particle size of coal dust on the explosibility of Pittsburgh seam bituminous coal as tested within LLEM. 49

Tables

Table 1. Summary of rock dusting requirements for various nations 6

Table 2. Average coal sizes from intake airways in mines in 10 MSHA Safety and Health Districts 16

Table 3. Average coal particle size from intake airways for various coal seams 17

Table A-1. Analyses of size of coal dust particles from intake airways in six MSHA District 2 mines 28

Table A-2. Analyses of size of coal dust particles from intake airways in seven MSHA District 3 mines ... 29

Table A-3. Analyses of size of coal dust particles from intake airways in seven MSHA District 4 mines ... 30

Table A-4. Analyses of size of coal dust particles from intake airways in six MSHA District 5 mines 31

Table A-5. Analyses of size of coal dust particles from intake airways in five MSHA District 6 mines 32

Table A-6. Analyses of size of coal dust particles from intake airways in five MSHA District 7 mines 33

Table A-7. Analyses of size of coal dust particles from intake airways in six MSHA District 8 mines 34

Table A-8. Analyses of size of coal dust particles from intake airways in seven MSHA District 9 mines ... 35

Table A-9. Analyses of size of coal dust particles from intake airways in five MSHA District 10 mines 36

Table A-10. Analyses of size of coal dust particles from intake airways in seven MSHA District 11 mines 37

Table B-1. Analyses of size of coal dust particles from return airways in 36 mines 40

Table B-2. Analyses of size of coal dust particles from return airways for seven Pittsburgh seam coal mines 42

Table C-1. Pittsburgh seam coal dust sizes 46

Table C-2. Limestone rock dust sizes 46

Table C-3. Average proximate and ultimate analyses of coal used in the LLEM experiments 47

Table C-4. LLEM inerting tests for Pittsburgh seam coal dust and limestone rock dust using a 40 ft long ignition zone 48

ACRONYMS AND ABBREVIATIONS USED IN THIS REPORT

% TIC	Percentage of Total Incombustible Content
ASTM	American Society for Testing and Materials
BEM	Bruceton Experimental Mine
CFR	Code of Federal Regulations
D_{med}	mass median diameter
D_s	surface mean diameter
D_w	mass or volume mean diameter
hvb	high volatile bituminous
hvCb	high volatile C bituminous
LLEM	Lake Lynn Experimental Mine
LLL	Lake Lynn Laboratory
LTA	low temperature ashing
lvb	low volatile bituminous
MSHA	Mine Safety and Health Administration
mvb	medium volatile bituminous
NIOSH	National Institute for Occupational Safety and Health
NP	non-propagation
OMSHR	Office of Mine Safety and Health Research
P	Propagation
PC	personal computer
TIC	Total Incombustible Content

UNIT OF MEASURE ABBREVIATIONS USED IN THIS REPORT

Btu/lb	British thermal unit per pound
cm	centimeter
ºC	degree Celsius
ft	foot
g/m^3	gram per cubic meter
hr	hour
kPa	kilopascal
m	meter
µm	micrometer or micron
mt/yr	million ton per year
ms	millisecond
%	percent
psi	pound-force per square inch
sec	second
ft^2	square foot
m^2	square meter

Dedication

This report was initially prepared by Kenneth L. Cashdollar and is dedicated to his memory. Ken passed away on March 4, 2009. Ken never wavered from his continuing commitment to conduct the highest quality, solution-oriented, scientific research focused on reducing the risk of explosion fatalities in the mining and chemical industries.

Recommendations for a New Rock Dusting Standard to Prevent Coal Dust Explosions in Intake Airways

Kenneth L. Cashdollar[1], Michael J. Sapko[2], Eric S. Weiss[3], Marcia L. Harris[4], Chi-Keung Man[5], Samuel P. Harteis[6], and Gregory M. Green[7]

Executive Summary

The workings of a bituminous coal mine produce explosive coal dust for which adding rock dust can reduce the potential for explosions. Accordingly, guidelines have been established by the Mine Safety and Health Administration (MSHA) about the relative proportion of rock dust that must be present in a mine's intake and return airways. Current MSHA regulations require that intake airways contain at least 65% incombustible content and return airways contain at least 80% incombustible content. The higher limit for return airways was set in large part because finer coal dust tends to collect in these airways. Based on extensive in-mine coal dust particle size surveys and large-scale explosion tests, the National Institute for Occupational Safety and Health (NIOSH) recommends a new standard of 80% total incombustible content (TIC) be required in the intake airways of bituminous coal mines in the absence of methane.

MSHA inspectors routinely monitor rock dust inerting efforts by collecting dust samples and measuring the percentage of TIC, which includes measurements of the moisture in the samples, the ash in the coal, and the rock dust. These regulations were based on two important findings: a survey of coal dust particle size that was performed in the 1920s, and large-scale explosion tests conducted in the U.S. Bureau of Mines' Bruceton Experimental Mine (BEM) using dust particles of that survey's size range to determine the amount of inerting material required to prevent explosion propagation.

Mining technology and practices have changed considerably since the 1920s, when the original coal dust particle survey was performed. Also, it has been conclusively shown that as the size of coal dust particles decreases, the explosion hazard increases. Given these factors, NIOSH and MSHA conducted a joint survey to determine the range of coal particle sizes found in dust samples collected from intake and return airways of U.S. coal mines. Results from this survey show that the coal dust found in mines today is much finer than in mines of the 1920s. This increase in fine dust is presumably due to the increase in mechanization.

In light of this recent comprehensive dust survey, NIOSH conducted additional large-scale explosion tests at the Lake Lynn Experimental Mine (LLEM) to determine the degree of rock dusting necessary to abate explosions. The tests used Pittsburgh seam coal dust blended as 38% minus 200 mesh and referred to as medium-sized dust. This medium-sized blend was used to

---

[1] Principal Physical Scientist (deceased), OMSHR, NIOSH, Pittsburgh, PA.
[2] Principal Physical Scientist (retired), OMSHR, NIOSH, Pittsburgh, PA.
[3] Senior Mining Engineer (Team Leader, Lake Lynn Laboratory Research and Mine Rescue Team), OMSHR, NIOSH, Pittsburgh, PA.
[4] Research Chemical Engineer, OMSHR, NIOSH, Pittsburgh, PA.
[5] Senior Service Fellow, OMSHR, NIOSH, Pittsburgh, PA.
[6] Lead Mining Engineer, OMSHR, NIOSH, Pittsburgh, PA.
[7] Physical Science Technician, OMSHR, NIOSH, Pittsburgh, PA.

represent the average of the finest coal particle size collected from the recent dust survey. Explosion tests indicate that medium-sized coal dust required 76.4% TIC to prevent explosion propagation. Even the coarse coal dust (20% minus 200 mesh or 75 μm), representative of samples obtained from mines in the 1920s, required approximately 70% TIC to be rendered inert in the larger LLEM, a level higher than the current regulation of 65% TIC.

Given the results of the extensive in-mine coal dust particle size surveys and large-scale explosion tests, NIOSH recommends a new standard of 80% TIC be required in the intake airways of bituminous coal mines in the absence of methane. The survey results indicate that in some cases there are no substantial differences between the coal dust particle size distributions in return and intake air courses in today's coal mines. The survey results indicate that the current requirement of 80% TIC in return airways is still appropriate in the absence of background methane.

Introduction

Despite the worldwide research on coal mine safety, coal mine explosions involving fatalities and injuries still occur [Dobroski et al. 1996; McKinney et al. 2002; Light et al. 2007]. Experimental studies by the Office of Mine Safety and Health Research[8] (OMSHR) and similar agencies in other countries have shown that inert rock dust acts as a heat sink, and mixing a sufficient quantity of inert rock dust with coal dust will prevent coal dust explosions [Cybulski 1975; Michelis et al. 1987, 1996; Reed et al. 1989; Lebecki 1991]. The U.S. mining law pertaining to rock dusting for the prevention of coal dust explosions was specified in the Federal Coal Mine Health and Safety Act of 1969 and was included in the Federal Mine Safety and Health Act of 1977 [U.S. Congress 1969 and 1977]. Current regulations are specified in Title 30, Part 75, Section 75.403 of the U.S. Code of Federal Regulations (CFR) [30 CFR[9] 2010]. Current regulations state that U.S. bituminous coal mines must maintain an incombustible content of at least 65% in the non-return (intake) airways and at least 80% in the return airways. Return airways require more inert material because there is greater risk of accumulation of finer coal dust. The U.S. regulations also require an additional 1.0% incombustible by weight for each 0.1% of methane in the ventilating air inside intakes and 0.4% additional incombustible for each 0.1% of methane in returns.

The total incombustible content (TIC) includes measurements of the moisture in the samples, the ash in the coal, and the rock dust. The 65% TIC required for intake airways was adopted based on the results of two studies. First, coal dust samples were collected and measured to determine the average size of coal dust particles. Next, full-scale experimental mine tests were conducted to determine the amount of rock dust required for coal particles of the size collected in the survey to be rendered inert [Nagy 1981]. The term "mine-size dust" was adopted in the mid-1920s and refers to coal dust that passes through a U.S. Standard 20-mesh sieve (850 μm) and contains 20% minus 200 mesh (75 μm). The justification for adopting this definition is given in Bureau of Mines Technical Paper 464 [Rice and Greenwald 1929]. Briefly, Technical Paper 464 indicates that coal dust samples collected from the mine floors had 5% to 40% of the material minus 200 mesh (75 μm) and that the values were weighted. For 80% of mines, the final values ranged from 15% to 25% through 200 mesh. Therefore, coal dust having 20% passing through 200 mesh was considered to be typical and termed "mine-size dust." The authors of Technical Paper 464 acknowledge that dust collected from ribs, roof, and timbers was finer, with 40% to 75% of the particles finer than 200 mesh, though they do not list the distribution of dust that would pass through sieves other than 200 mesh. Also missing from the report are details on the total number of mines surveyed and the total number of samples analyzed for coal particle size. Many years later, Public Law 552 (82nd Congress, 1952) required 65% incombustible content for most mines entries but it did not differentiate between intake and return areas.

The quantities of rock dust required in the return airways in bituminous coal mines in the United States were increased to 80% by enactment of Public Law 91–173, the Federal Coal Mine Health and Safety Act of 1969. Section 304(a) mandated that coal dust shall be cleaned up and not permitted to accumulate in active workings or on electrical equipment. Paragraph (b) noted that when excessive dust is raised, water, water plus a wetting agent, or other no less effective

---

[8] The Pittsburgh Research Center was part of the U.S. Bureau of Mines until 1996, when it was transferred to the National Institute for Occupational Safety and Health (NIOSH) and became known as the Pittsburgh Research Laboratory. Since 2009, it is referred to as OMSHR.
[9] *Code of Federal Regulations.* See CFR in references.

agent shall be applied to abate dust, especially in distances less than 40 feet from the face to minimize explosion hazards. Paragraph (c) required that all underground areas where the incombustible content is too low shall be rock dusted to within 40 feet of the face. All crosscuts that are less than 40 feet from a working face shall also be rock dusted. Section 304(d) reads as follows:

> Where rock dust is required to be applied, it shall be distributed upon the top, floor, and sides of all underground areas of a coal mine and maintained in such quantities that the incombustible content of the combined coal dust, rock dust, and other dust shall be not less than 65 per centum, but the incombustible content in the return air courses shall be no less than 80 per centum. Where methane is present in any ventilating current, the per centum of incombustible of such combined dusts shall be increased 1.0 and 0.4 per centum for each 0.1 per centum of methane, where 65 and 80 per centum respectively, of incombustibles are required.

The aforementioned requirement of 80% TIC in return airways represents an increase over previous standards for return airways. The entire standard was based on earlier research with "mine-size dust." The incombustible content needed to prevent propagation given a particular coal dust size is also dependent, to a lesser extent, on the volatility content of the coal. The decision to require all coal dusts except anthracite to have 65% TIC was made in 1927 by the Mine Safety Board. Decision No. 5, relating to rock dusting [Rice 1927], was superseded and clarified by Decision No. 32 [Mine Safety Board, 1937]. All Federal mine codes and laws since the mid-1920s have contained the same requirement. The requirement to have a 65% incombustible content for all coals except anthracite was made to simplify rock dusting practices. Coals that have a volatile ratio [volatile ratio = volatile content / (volatile content + fixed carbon)] of less than 0.2 provide a greater margin of explosion protection than coals having a volatile ratio higher than 0.2 [Nagy 1981].

The effect of coal particle size on explosibility is illustrated in Figure 1 as adapted from Rice et al. [1922] and Rice and Greenwald [1929]. This figure shows the amount of incombustible dust required to prevent propagation of an explosion for Pittsburgh high volatile bituminous coal dust with 10% to 80% passing through a 200 mesh (75 µm) sieve. Each of the data points is an individual explosion test conducted in the NIOSH-OMSHR Bruceton Experimental Mine (BEM). The curve is the boundary between mixtures that can propagate an explosion (below line) and mixtures that cannot propagate an explosion (above line). These data were used to support the 65% incombustible requirement for intake and return airways based on "mine-size dust" of the time.

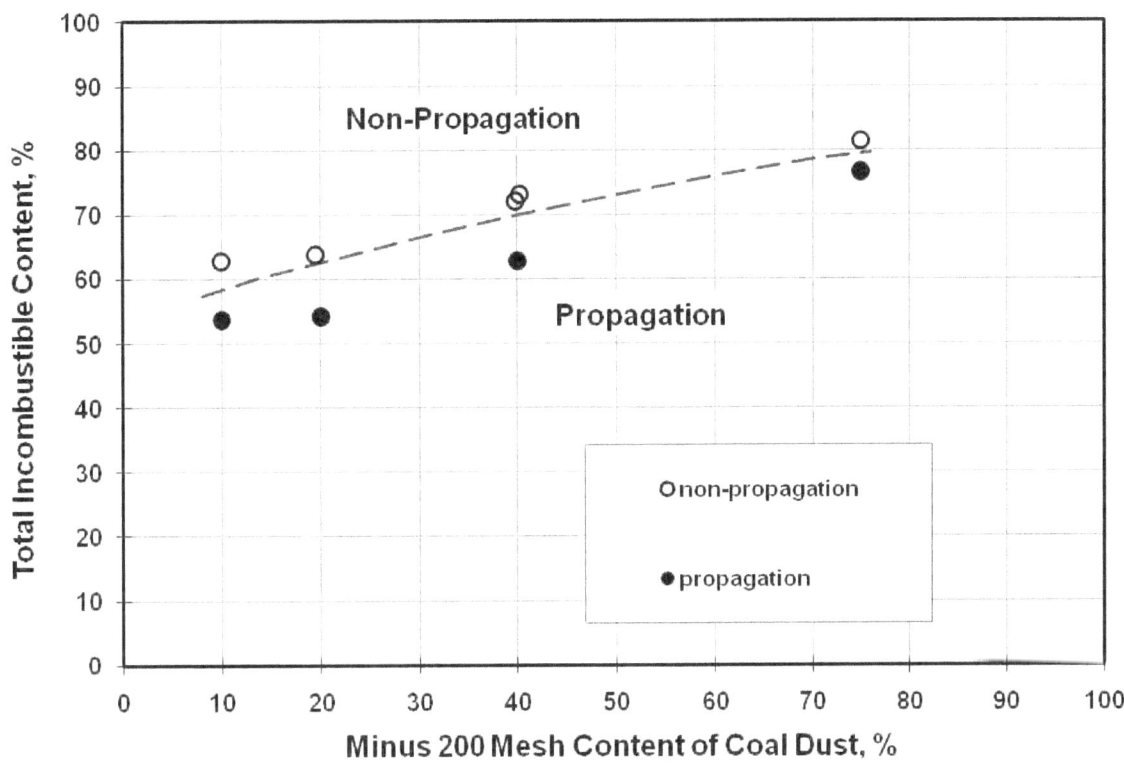

Figure 1. Effect of particle size of coal dust on the explosibility of Pittsburgh seam bituminous coal as tested within BEM.

Comparison of International Rock Dusting Requirements

Rock dust has been used for about 100 years as a precautionary measure to protect against dust explosions. It is generally agreed that the effectiveness of rock dust lies in its ability to be simultaneously dispersed with coal dust, and, by serving as a heat sink, thus prevent flame propagation. Most leading coal-producing nations have similar requirements, some more stringent and some less stringent than those enforced in the United States. A partial listing of these requirements is given in Table 1. Passive barriers have been deployed in most leading coal-producing nations to provide supplemental protection against coal dust explosions. Conveyor belt entries have received emphasis. Barriers are designed to quench an explosion immediately on arrival at the location [Cybulski 1975, Liebman et al 1974, and Sapko et al 1989].

Table 1. Summary of rock dusting requirements for various nations

Country	TIC %	Volatile matter %	Methane %	Comments
Australia Queensland	85–80 (return)	—		85% TIC ≤ 200 m from the face
				80% TIC > 200 m from the face
	85–70 (intake)	—		85% TIC ≤ 200 m from the face
				70% TIC > 200 m from the face
				Supplemental protection—barriers
Australia NSW	85–70 (return)	—		85% TIC ≤ 200 m from the face
				70% TIC > 200 m from the face
	80–70 (intake)	—		80% TIC ≤ 200 m from the face
				70% TIC > 200 m from the face
				Supplemental protection—barriers
Canada (Nova Scotia)	75 (intake)	—	<1	
	80 (return)	—	>1	
Czech Republic	80 (intake/return)	—	<1	Supplemental protection—barriers
	85 (intake/return)	—	>1	
Slovakia	80 (intake/return)	—	<1	Supplemental protection—barriers
	85 (intake/return)	—	>1	
Germany	80 (intake/return)	—		Supplemental protection—barriers
Japan	78 (intake/return)	35	<1	Specific requirements depend on ash, moisture and volatile content, the gassiness of the seam, and the fineness of the rock dust used.
	83 (intake/return)	35	>1	
Poland	70 (intake/return)	>10		70% in "non-gassy" roadways
		>10		80% in "gassy" roadways
				Supplemental protection—barriers
South Africa	80 (intake)	—		80 % TIC ≤ 200 m from the face
				65% TIC > 200 m from the face
	80 (return)	—		80% TIC for 1000 m from the face
				Supplemental protection—barriers
United Kingdom	50 (intake/return)	20		Supplemental protection—barriers
	65 (intake/return)	27		
	72 (intake/return)	35		
	75 (intake/return)	>35		
United States	65 (intake)	—	1.0 / 0.1	Add 1% TIC / 0.1% methane
	80 (return)	—	0.4 / 0.1	Add 0.4% TIC / 0.1% methane

From 1985 through 2001, numerous coal dust explosion tests were conducted in the single entry D-drift at LLEM to determine the concentration of rock dust required to prevent explosion propagation of samples with varying coal dust particle sizes, volatilities, mine entry size, and other related properties. The LLEM drifts (20-ft or 6-m wide by 6.5-ft or 2-m high) are more representative of current U.S. underground coal mine geometries compared to the much smaller BEM entries (9-ft or 2.7-m wide by 6-ft or 1.8-m high).

The factors that can influence the amount of admixed rock dust required to make coal dust inert include coal and rock dust particle size distribution, coal dust volatile content, and the additional presence of methane. Much knowledge has been obtained from experimental mine and laboratory dust explosion research during the past 3 decades. Investigators have examined the effects of rock dust inerting requirements, the minimum explosible coal dust concentrations, the effect of volatile matter on the explosibility of coal dusts, the effect of the size of coal and rock dust particles, and the effect of background methane in full-scale experimental mines and in laboratory test vessels [Sapko et al. 1987a, b; 1989; 1998; 2000; Cashdollar 1996; Cashdollar and Hertzberg 1989; Cashdollar and Chatrathi 1993; Cashdollar et al. 1987; 1988; 1992a, b, c; 2007]. Further research evaluated the effects of pulverized versus coarse coal particle size [Weiss et al. 1989], coal volatility, extinguishment, and pyrolysis mechanisms [Hertzberg et al. 1987; 1988a, b; Conti et al. 1991; Greninger et al. 1991]. The clear cumulative consensus of these studies is that dust particle size emerges as the single most influential factor controlling coal dust explosion propagation. Therefore, the primary focus of this research was to examine the effect of coal particle size of Pittsburgh coal while holding other factors constant.

To determine compliance with current regulations, inspectors from MSHA periodically collect samples of deposited dust from various areas in a mine. The MSHA laboratory determines TIC and compares it with the TIC requirement. This TIC requirement is based on a mean coal particle size of 20% minus 200 mesh and assumed to be constant throughout the intake entries. The size of the coal dust component is not measured by MSHA laboratories as part of the explosibility assessment.

This report presents the results of extensive in-mine coal dust particle size surveys of dust samples collected from intake airways in 61 U.S. coal mines, representing all 10 MSHA bituminous Coal Mine Safety and Health Districts (Figure 2). MSHA District 1 covers anthracite mines in Pennsylvania, which do not require rock dusting. A preliminary version of this research with data from 50 mines was published by Sapko et al. [2007]. Samples from return airways in 36 mines were also size analyzed. A series of large-scale dust explosion tests was then conducted at the LLEM using the average of the finest coal particle size from the MSHA district intake survey results to determine the incombustible content necessary to prevent explosion propagation.

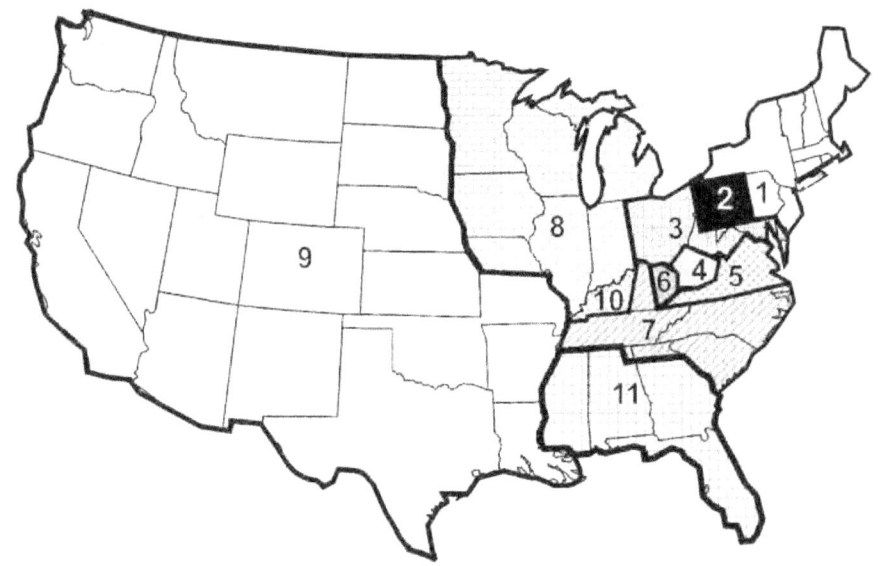

Figure 2. MSHA Coal Mine Safety and Health Districts, identified by number.

Experimental Procedures

To assess current variations in coal particle size from various underground coal mining operations, MSHA coordinated the acquisition of mine dust samples from the 10 bituminous Coal Mine Safety and Health Districts. The dust samples were among those routinely collected by mine inspectors to assess compliance with 30 CFR 75.403. The detailed sampling protocols are summarized in the General Coal Mine Inspection Procedures and Inspection Tracking System [MSHA 2008]. The samples were sent to the MSHA laboratory at Mt. Hope, WV, and analyzed for total incombustible content (TIC). The TIC includes measurements of the moisture in the samples, the ash in the coal, and the rock dust. The incombustible analysis procedure [Montgomery 2005] begins by passing the sample through a 20-mesh sieve (850 μm) and then oven drying the minus 20-mesh material for 1 hr at 105°C. The weight lost during drying constitutes the as-received-moisture in the sample. Next, the dried sample is heated in an oven that is ramped up over 1.5 hr and held at 515°C for about 2.5 hr to burn off the combustible coal fraction, thereby leaving the ash and incombustible material. This low temperature ashing (LTA) burns off the coal but does not decompose the limestone rock dust. The amount of the remaining ash material plus the as-received-moisture divided by the initial weight is reported as %TIC. Portions of each dust sample that were not needed for TIC measurement were sent to NIOSH-OMSHR for the analyses of coal particle sizes.

At OMSHR, the limestone (or marble) rock dust was leached from the sample using hydrochloric acid. In this leaching method used in the laboratory, dilute hydrochloric acid was added to the dust sample in a beaker and heated on a hotplate. The acid reacted with the limestone or marble rock dust, producing foam while releasing carbon dioxide. Sufficient acid was added until all foaming stopped. The hotplate kept the slurry near its boiling point for about 1 hr. After the slurry cooled, the acid-insoluble residue was filtered from the acid. The solid residue was rinsed with water and isopropanol and then transferred to a large evaporating dish. The residue was dried at 110°C for 3 hr. Agglomerates were broken with a spatula. The residue consisted of coal plus other insoluble mineral matter.

The dried residue was then classified into the different size fractions using a sonic sieve, which provided particle separation by combining two motions—a vertical oscillating column of air, and a repetitive mechanical pulse. Occasionally the tops of the sieves were brushed to break up any remaining agglomerates. The sieves are 8 cm in diameter and include the following sizes: 20 mesh (850 µm), 30 mesh (600 µm), 40 mesh (425 µm), 50 mesh (300 µm), 70 mesh (212 µm), 100 mesh (150 µm), 140 mesh (106 µm), 200 mesh (75 µm), 270 mesh (53 µm), and 400 mesh (38 µm). After the sieving was completed, the weight of sample on each sieve was recorded.

Because the residue from the leaching process contained other inert mineral matter that did not react with the acid, a correction to the size analysis had to be made. First, the residue was grouped into three size fractions: minus 200 mesh, 200 - 70 mesh, and plus 70 mesh. At OMSHR, these three fractions were heated to 515°C to determine the incombustible or non-coal content, using an LTA method similar to that of the MSHA laboratory at Mt. Hope. The analyses of sieve size were then corrected for the non-coal content (insoluble mineral matter) in the three size groupings. The amount of this insoluble mineral matter in the samples varied greatly, but it was generally in the 20% to 50% range. For most of the samples analyzed, the insoluble mineral matter was finer than the coal particles. Therefore, after correction for the mineral matter, the corrected minus 200-mesh amount would be less than the original minus 200-mesh amount determined by sonic sieving alone. There was a wide range of correction values, but a value of 39% minus 200 mesh from the original sieving data might typically be reduced to ~31% minus 200 mesh after correcting for the mineral matter. Details of the size analyses, listing both original and corrected data, are included in the tables of Appendixes A and B.

The total size analysis procedure (acid leaching, sieving, and correction for remaining incombustible matter) was verified by using prepared mixtures of coal and rock dust. First, the particle size distribution of the coal sample was determined by sieving. Next, samples of coal and rock dust were mixed together, and the rock dust was leached from the mixture. The residue was then sieved and corrected via LTA for any remaining incombustible matter in the size fractions. Data for a mixture of 30% medium-sized Pittsburgh seam high volatile coal and 70% limestone rock dust are shown in Figure 3. Both the cumulative and differential size distributions (by mass) are shown. A gold dashed vertical line shows the 200 mesh (75 µm) size and a dot-dashed vertical orange line shows the 70 mesh (212 µm) size. Both the original coal (red data curves) and acid-leached residue from the mixture (blue data curves) had their size analyses corrected via LTA for any remaining incombustible matter. For this mixture, both the percentage through 200 mesh and the median size (50% point on the cumulative distribution curve) were almost identical for the original coal and the residue from the acid-leached mixture. Figure 4 shows similar data for a mixture of 30% medium-sized Pittsburgh seam coal, 60% limestone rock dust, and 10% kaolin clay (to simulate possible shale dust in the sample). The original coal data are shown by the red curves and the acid-leached residue data from the mixture are shown by the blue curves. Figure 4 also shows close agreement for the percentage through 200 mesh and almost identical median values from the two cumulative curves. Original and acid-leached Blue Creek seam and Pocahontas seam samples were compared, but without any added rock dust. In general, the size analyses after leaching were within 1% to 3% of the amount of minus 200 mesh material (data not shown). Therefore, there is no evidence that the acid-leaching procedure compromises the accuracy of the sieve analysis of the coal dust.

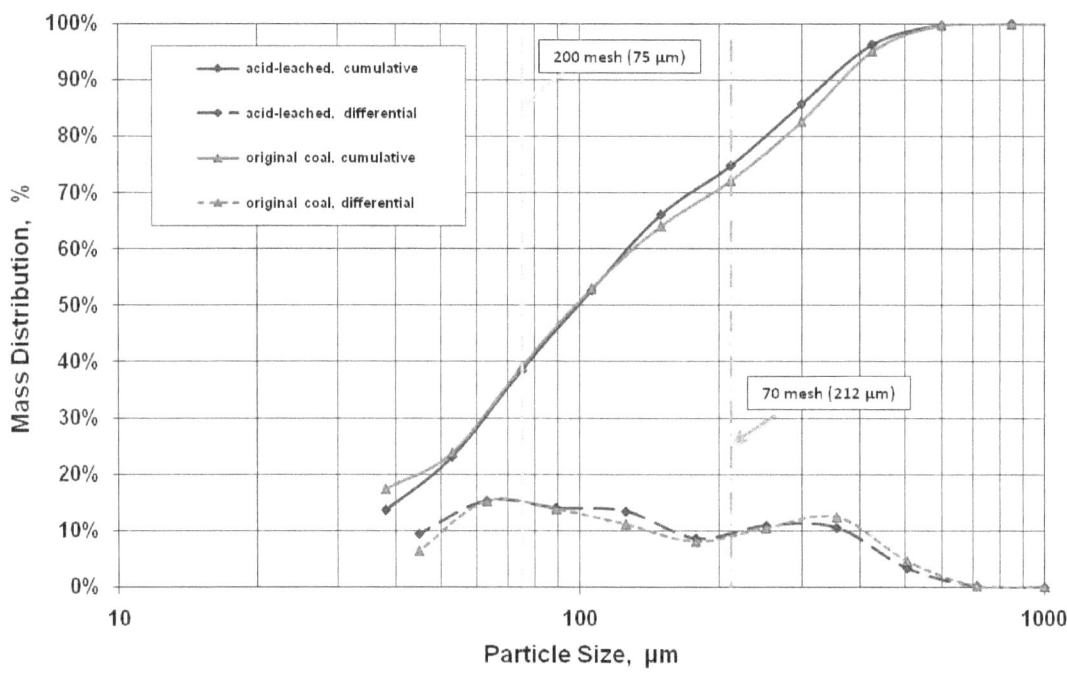

Figure 3. Original analyses of coal sieve size and analyses of sieve size of acid-leached mixture containing 30% medium-sized Pittsburgh coal and 70% limestone rock dust.

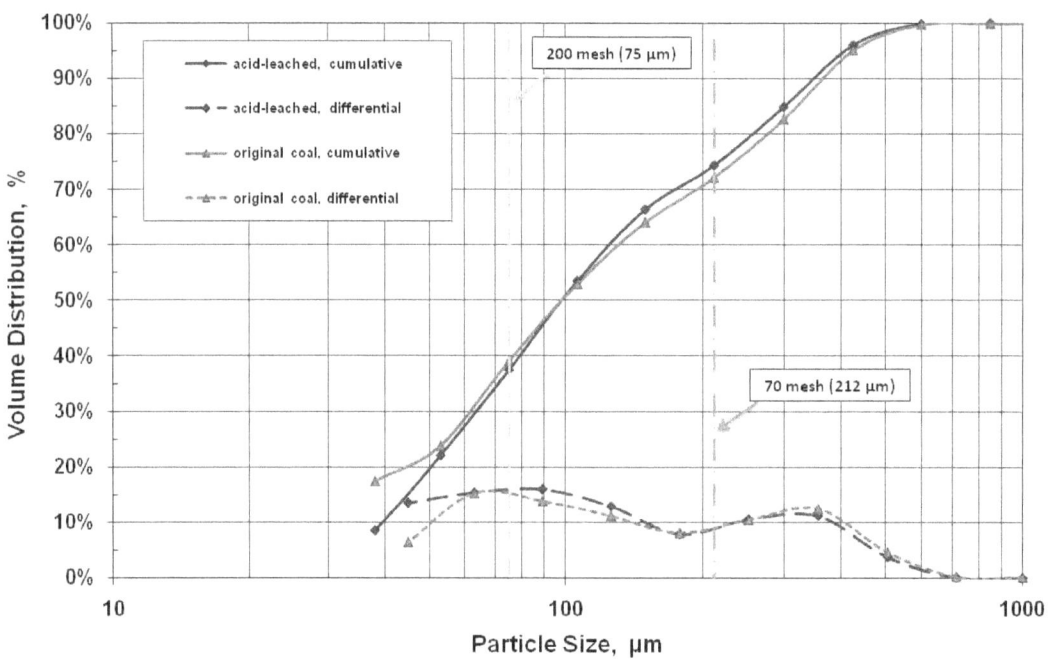

Figure 4. Original analyses of coal sieve size and analyses of sieve size of acid-leached mixture containing 30% medium-sized Pittsburgh coal, 60% limestone rock dust, and 10% kaolin clay.

The large-scale explosion tests were conducted in the LLEM, which is shown in the plan view of Figure 5 [Triebsch and Sapko 1990]. This is a former limestone mine, and five new drifts (horizontal passageways in a mine) were developed to simulate the geometries of modern U.S. coal mines. The mine has four parallel drifts—A, B, C, and D. D-drift is a 1,640-ft-long (500-m) entry that can be separated from E-drift by an explosion-resistant bulkhead door. In order to simulate room and pillar workings, drifts A, B, and C can be used. These three drifts are approximately 1,600 ft long (490-m), with seven crosscuts at the inby end. Drifts C and D are connected by E-drift, a 500-ft-long (152-m) entry that simulates a longwall face. Explosion tests can be conducted in the single entry D-drift, the multiple entry area of A-, B-, and C-drifts, or various other configurations including the longwall E-drift. The entries are about 20 ft wide (6-m) by about 6.5 ft high (2-m), with cross-sectional areas of 130–140 ft^2 (12–13 m^2). The LLEM bulkhead door and some of the other infrastructure were designed to withstand explosion overpressures of up to 100 psi (7 bar or 700 kPa). Higher pressures have been recorded at areas away from these structures. Previous publications described the LLEM coal dust explosion test procedures and the results of LLEM explosion research and post-explosion observations [Weiss et al. 1989; Greninger et al. 1991; Cashdollar et al. 1992b, c; Sapko et al. 1998; 2000].

Each LLEM drift has 10 data-gathering stations inset in the rib, which houses a strain gauge transducer to measure the explosion pressure and an optical sensor to detect flame arrival. The wall pressure is perpendicular to the gas flow and is the pressure that is exerted in all directions. This quasi-static pressure is called the "static pressure" by Nagy [1981, p. 58] to differentiate it from the dynamic pressure, although the "static pressure" does vary with time during the explosion. The dynamic or wind pressure is directional. The total explosion pressure is the sum of the quasi-static pressure and the wind or dynamic pressure. Other instruments such as dynamic pressure sensors, heat flux gauges to measure explosion temperatures, optical probes to measure dust dispersion, and video cameras may be installed at various locations in the LLEM. During the explosion tests, a PC-based National Instruments data acquisition system collected the data from the various instruments at a sampling rate of 1,500 to 5,000 samples per second.

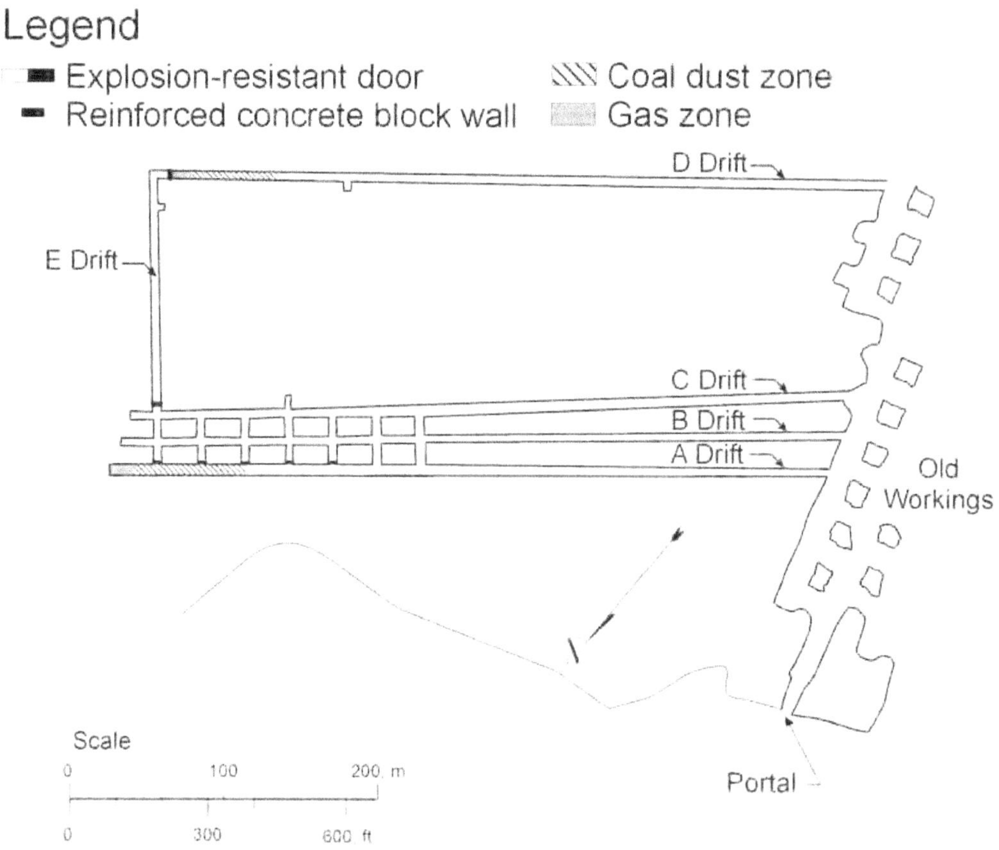

Legend
- Explosion-resistant door
- Reinforced concrete block wall
- Coal dust zone
- Gas zone

Figure 5. Plan view of the Lake Lynn Experimental Mine (LLEM).

The LLEM dust explosion tests, described in this paper, were conducted in D-drift and more recently in a modified single entry section of A-drift. These drifts were isolated from E-drift by means of the explosion-resistant movable bulkhead doors (Figure 5). The tested coal dusts were prepared in the NIOSH coal grinding and pulverizing facilities located at the OMSHR facility at Bruceton. The coal and rock dust particle size data used in the LLEM explosion studies from the mid-1980s through 2008 are presented in Appendix C: Table C-1 and Table C-2, and coal analysis is presented in Table C-3. The size distributions of the limestone from the 1980s and from 2007 are similar, so comparisons of explosion inerting results from these periods are valid. The typical D-drift dust explosion test ignition zone (Figure 6) was located in the first 40-ft (12-m) as measured from the face (closed end). This 10% methane air zone was ignited by electric matches. In the rock dust inerting tests, the coal dust and limestone rock dust mixture was placed half on roof shelves made of expanded polystyrene and half on the floor as illustrated in Figure 7 and Figure 8. These roof shelves were suspended 1.5 ft (0.5 m) from the mine roof on 10-ft (3-m) increments throughout the dust zone. This dust distribution technique, developed through extensive testing at BEM and LLEM, is used to enable reproducibility of experimental conditions. The length of the dust zones during these inerting tests in D-drift varied as follows: 210, 270, 390, 420, 450, and 600 ft long (64, 82, 119, 128, 137, and 183 m). These dust zones started just outby the end of the 40-ft-long ignition zone, that is, the 210-ft-long dust zone extended from 40 to 250 ft (12 m to 76 m) as measured from the face. Although the majority of the dust zones were 210 ft long, the longer dust zones were used for several reasons that differed depending on the experiment. The extension of flame travel through and beyond the longer dust zones for a particular incombustible content was always compared to a similar 210-ft-long dust zone to verify that the flame propagation was not being overdriven by the methane ignition zone (which would typically travel ~200 ft or ~61 m from the closed end). Non-propagation is defined as no sustained flame propagation of the dust mixture. Propagation is defined as flame propagation of the dust mixture.

The nominal dust loading reported for the LLEM tests assumes that all of the dust was dispersed uniformly throughout the cross-section. For the LLEM tests, the test drift was thoroughly washed down after each test. Dehumidified air was passed through the entry, and the entry was allowed to dry several days before dust was loaded for the next test.

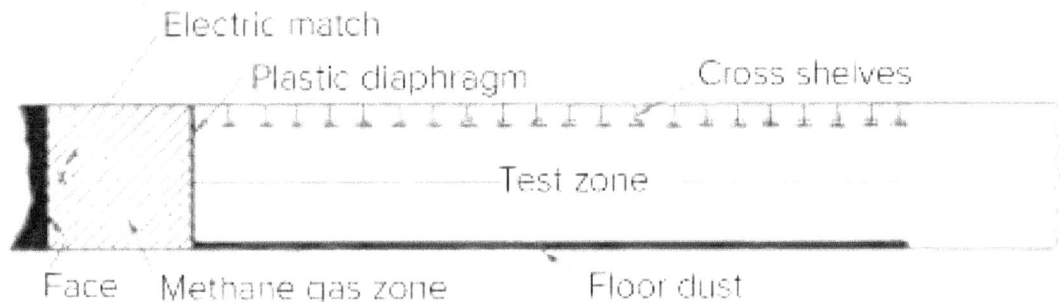

Figure 6. Side view of LLEM A-drift and D-drift test zones for determining rock dust inerting requirements.

Figure 7. Placing coal and rock dust mixture on shelves in the LLEM.

Figure 8. Distributing test dust mixture at the LLEM.

Size Data for Intake Airways

For this study, a total of 217 samples of mine dust from intake airways of 61 coal mines in the 10 MSHA bituminous districts were analyzed for particle size. For each mine, samples were usually collected from two or more entries. For most analyses, multiple samples from a mine entry were combined to give an average size distribution for that entry. Most of the samples were band samples, also known as perimeter samples, but some were floor and rib samples, floor and roof samples, or floor-only samples [MSHA, 2008, p. 60]. The detailed size data for each sample and each mine are listed in the tables of Appendix A. The mines are identified only as A, B, C, etc., so that the individual mines remain anonymous. Columns three and four of the tables in Appendix A list the incombustible percentage (from the MSHA Mt. Hope Laboratory) and the soluble in acid percentage, as measured at NIOSH-OMSHR. Columns five and six of the tables list the original size analyses. Column seven lists the weighted average of the ash or incombustible fraction of the acid-leached material. The remaining columns list the corrected size analyses. Table 2 lists the summary intake coal dust size data by the MSHA Coal Mine Safety and Health District. Column two lists the states within each MSHA District from which samples were obtained. There may be additional states within some districts from which there were no samples obtained. Columns three and four of the table list the number of mines and total number of combined samples per district. Columns five through twelve list the average percentage through the various sieves. The column for minus 200 mesh (75 µm) lists both the average value and the associated standard deviation. The standard deviations for the other sieve values are listed in the tables of Appendix A. The last column lists the average and standard deviation for the mass median particle diameter (50% point on the cumulative distribution curve), which was interpolated from the corrected sieving data. The cumulative size data for MSHA Districts 3, 9, and 11 are shown in Figure 9. MSHA District 11 has the finest dust, with 37% minus 200 mesh, and the western states (District 9) have the coarsest dust, with 27% minus 200 mesh. District 3 (northern WV, OH, and MD) has an intermediate size. The averages for all MSHA Districts are 31% minus 200 mesh, 61% minus 70 mesh, and a mass median particle diameter of ~156 µm. This is finer than particles measured in the 1920s.

Table 2. Average coal sizes from intake airways in mines in 10 MSHA Safety and Health Districts

District	States	Mines	Samples	-270 mesh or < 53 µm, %	-200 mesh or < 75 µm, %	-140 mesh or < 106 µm, %	-100 mesh or < 150 µm, %	-70 mesh or < 212 µm, %	-50 mesh or < 300 µm, %	-40 mesh or < 425 µm, %	-30 mesh or < 600 µm, %	Dmed, µm
2	PA	6	20	23	29 ± 4	37	47	59	72	85	95	165 ± 27
3	OH, MD, No. WV	7	22	26	33 ± 9	41	51	62	74	87	96	149 ± 42
4	So. WV	7	23	25	30 ± 6	38	48	60	73	87	97	165 ± 39
5	VA	6	20	25	31 ± 8	40	50	62	74	86	96	157 ± 36
6	Eastern KY	5	24	25	31 ± 7	39	49	59	72	85	96	160 ± 37
7	Central KY	5	19	29	34 ± 10	43	53	62	74	86	95	140 ± 48
8	IN, IL	6	18	24	29 ± 5	37	47	57	71	85	96	170 ± 31
9	CO, NM, UT	7	20	21	27 ± 3	36	46	57	71	85	96	172 ± 26
10	Western KY	5	28	23	29 ± 4	39	50	61	74	86	96	152 ± 24
11	AL	7	23	30	37 ± 10	48	60	73	84	92	97	128 ± 46
10 Districts Average			217	25	31	40	50	61	74	86	96	156

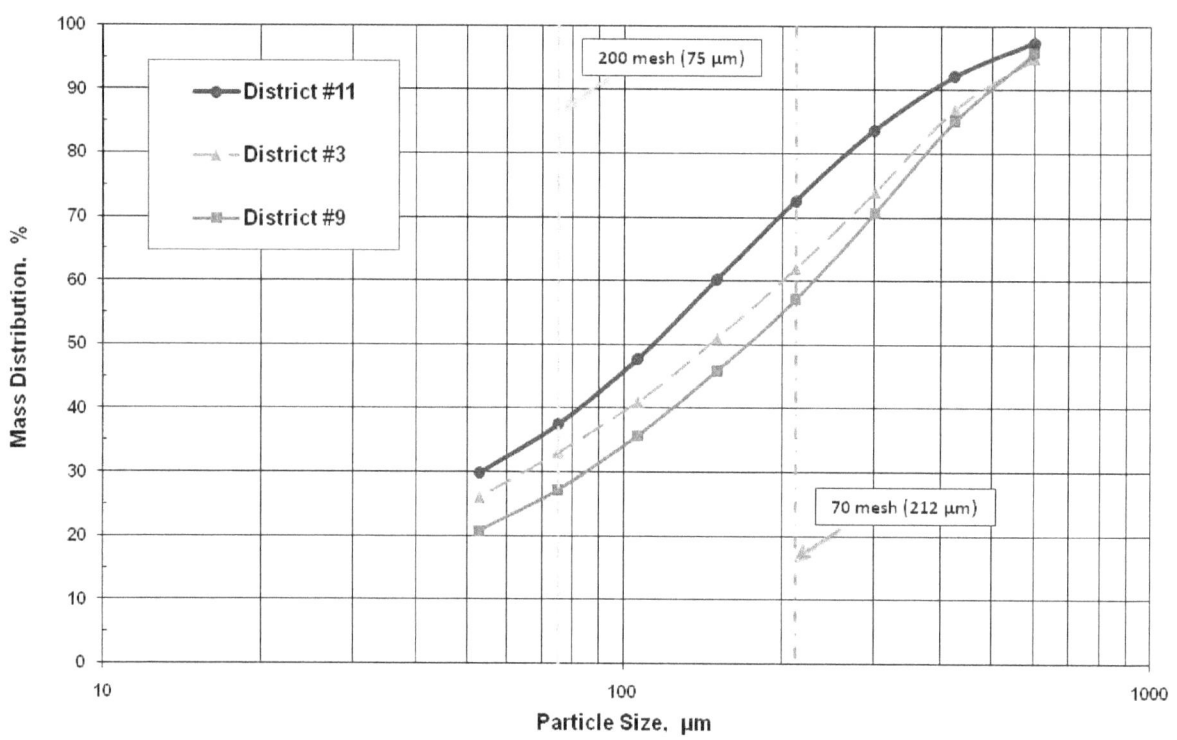

Figure 9. Coal particle size by MSHA district.

Table 3 lists the average coal dust particle sizes for intake airways for various coal seams or groups of adjacent coal seams. The eastern bituminous coal seams are those in the Appalachian Mountains from Pennsylvania to Alabama. Only the seams that included samples from two or more mines are listed. The coal rank is also listed in the first column, with hvb, mvb, and lvb indicating high, medium, and low volatile bituminous coal, respectively [ASTM 2008]. The mid-eastern seams are those in Illinois, Indiana, and western Kentucky. These seams are known by different names in different states, as listed in the table. The western coal seams include various high volatile C bituminous (hvCb) coals in Colorado or Utah. The coal samples from the Hazard #4 seam in Kentucky and the Blue Creek seam in Alabama are the finest, with 40% of the samples less than 200 mesh. However, the Hazard seam data are based on samples from only two mines and may not represent the area as well as the Blue Creek seam data. The Pittsburgh seam coal in OH, PA, and WV has 32% minus 200 mesh. The cumulative size data for the Blue Creek, Pittsburgh, and Herrin coal seams are shown in Figure 10.

Table 3. Average coal particle size from intake airways for various coal seams

Coal Seams	States	Mines	Samples	−270 mesh or < 53 µm, %	−200 mesh or < 75 µm, %	−140 mesh or < 106 µm, %	−100 mesh or < 150 µm, %	−70 mesh or < 212 µm, %	−50 mesh or < 300 µm, %	−40 mesh or < 425 µm, %	D_{med}, µm
Eastern Bituminous Coal Seams											
Pittsburgh, hvb	PA, OH, WV	9	36	25	32 ± 7	40	50	62	74	87	152 ± 34
Upper or Lower Kittanning, hvb	PA, WV	3	6	20	27 ± 7	34	43	54	67	82	187 ± 42
Eagle, hvb	WV	2	5	20	25 ± 7	33	44	56	70	85	187 ± 44
Powellton, hvb	WV	2	7	24	28 ± 5	36	45	56	69	84	180 ± 36
Pocahontas #3 & #5, lvb	WV, VA	3	11	26	32 ± 6	40	50	61	73	86	154 ± 36
Raven, hvb	VA	2	6	27	35 ± 10	45	57	70	80	89	138 ± 44
Alma, Cedar Grove, Darby, Upper Elkhorn #1 or #3, hvb	KY	5	25	27	32 ± 7	40	50	60	73	86	154 ± 38
Hazard #4, hvb	KY	2	8	34	40 ± 12	49	60	69	78	87	105 ± 43
Pratt coal seam, hvb	AL	2	6	25	31 ± 5	40	51	63	77	89	155 ± 34
Blue Creek coal seam, mvb	AL	5	17	31	40 ± 10	50	63	76	86	93	119 ± 47
Mid-Eastern Bituminous Coal Seams											
Springfield, Illinois #5, or W. Kentucky #9	KY, IL, IN	5	20	24	30 ± 5	39	50	61	74	87	155 ± 29
Herrin, Illinois #6, or W. Kentucky #11	KY, IL	4	14	21	27 ± 3	36	47	58	71	84	167 ± 25
Western Bituminous Coal Seams											
various hvCb seams in Colorado	CO	4	9	21	27 ± 3	36	46	57	70	84	174 ± 29
various hvCb seams in Utah	UT	2	4	18	25 ± 3	33	44	58	73	88	177 ± 15

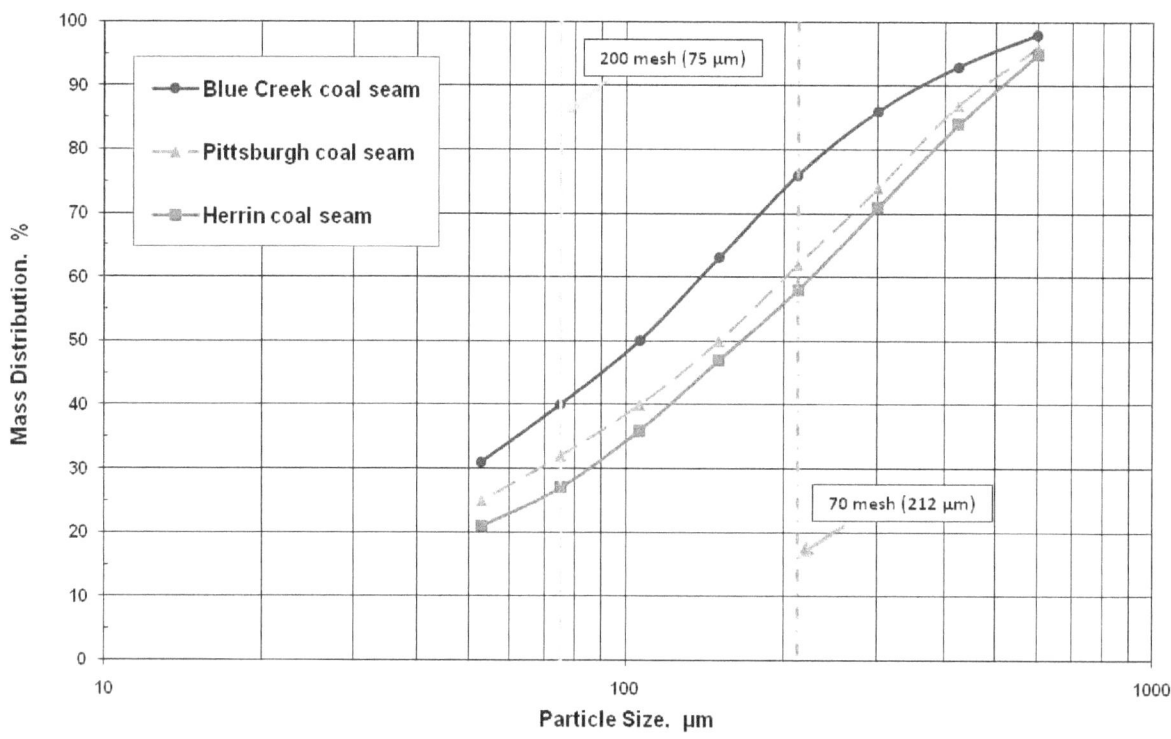

Figure 10. Coal particle size by coal seam.

Size Data for Return Airways

For this study, a total of 44 samples of mine dust was taken from return airways of 36 coal mines in the 10 MSHA bituminous districts and were analyzed for particle size. Samples were collected from one or more entries in each mine. Similar to the intake airways, multiple samples from a mine entry were combined to give an average size distribution for that entry. Most of the samples were band samples, but some were floor and rib samples, floor and roof samples, or floor-only samples. The detailed size data for the return airways are listed in Table B-1 in Appendix B. For the returns, there was a much larger variation in the coal dust size. Many samples had percentages of minus 200 mesh dust, which were similar to those of the intake samples. However, 8 of the 44 samples had 60% to more than 80% minus 200 mesh. The only coal seam for which there were sufficient samples to calculate a representative average size was the Pittsburgh coal seam. The coal samples had an average of 62% minus 200 mesh (Table B-2 in Appendix B), finer than the intake coal samples from the Pittsburgh seam.

MSHA Dust Survey Results from Intake and Return Airways

MSHA, from January 2005 to February 2008, collected and determined the TIC for 65,536 intake and 60,663 return airway samples from underground coal mines. Each dust sample represents about 500 ft (152 m) of mine entry. The intake airways are currently required to contain at least 65% TIC. Approximately 87% contained ≥ 65% TIC, while ~13% contained < 65% TIC and thus were non-compliant. The fact that ~13% of the samples collected were found to be non-compliant illustrates the scope of the problem. Considering that each sample may represent up to 500 ft (152 m) of mine entry, these ~13%, or 8,323 samples, represent more than 788 miles (1,268 km) of underground coal mine entries that were deficient. At the other extreme, 66% of the intake samples contained more than 80% TIC, and ~54% contained more than 85% TIC. This indicates that rock dusting efforts exceed requirements in a majority of samples, because the average TIC among all samples was ~82% TIC.

A similar TIC distribution is observed for return airway samples. Current MSHA regulations require 80% TIC for return airways. Analysis of 60,663 samples revealed that ~72% of samples contained ≥ 80% TIC while ~28% contained < 80% TIC. The average TIC for return samples was 85%, which is ~3% higher than the intake average of ~82%.

The MSHA dust survey data indicate that many areas have more than sufficient inert material. However, there are still a large number of areas where rock dusting efforts are insufficient to prevent coal dust explosions.

Limestone Rock Dust Inerting

Prior to having recent access to the MSHA band samples collected from underground coal mines throughout the United States, there was growing evidence from limited dust surveys that the coal dust particle size had been decreasing since the promulgation of the existing rock dusting regulations. This decrease occurred as new mining technologies were adopted by the industry. Numerous coal dust explosion tests have been conducted in the LLEM to specifically quantify the concentration of rock dust required to prevent propagation of a high volatile coal as a function of coal dust particle size. Table C-4 shows a composite of these experiments. Details of these experiments can be found in Table C-4 in Appendix C along with a discussion highlighting the specific experimental results.

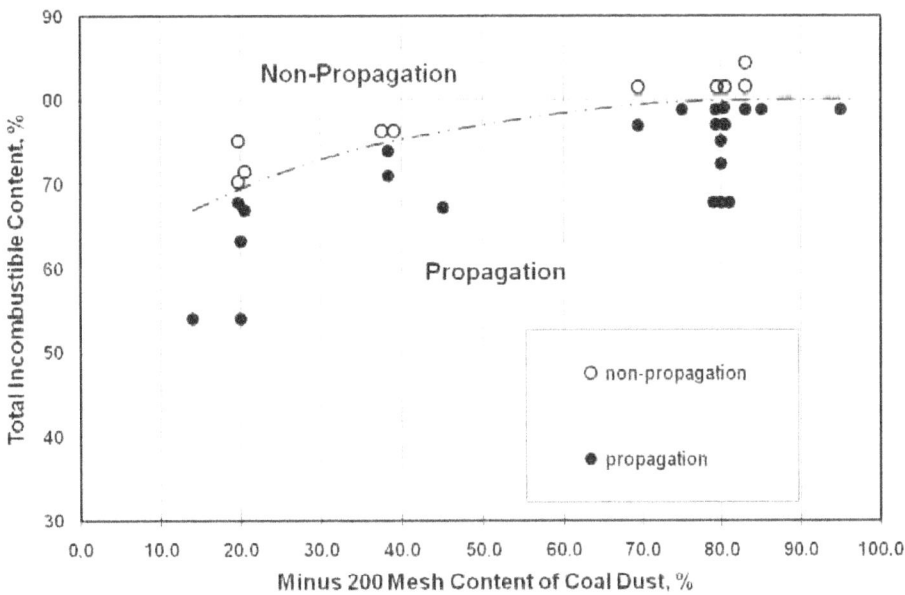

Figure 11. Effect of particle size of coal dust on the explosibility of Pittsburgh seam bituminous coal as tested within LLEM.

Following the coal dust survey, additional large-scale explosion experiments were conducted using medium-sized dust (38% minus 200 mesh or 75 microns—Table C-1) to better define the boundary between explosion propagation and non-propagation. Medium-sized dust was formulated with a blend of 2008 pulverized and 2008 coarse dust (Table C-1) of Pittsburgh seam coal to represent the average of the finer dusts collected from the survey. However, approximately 12% of the collected intake airway dust samples (26 of the 217 samples) ranged in size from 39 to 63% minus 200 mesh. These finer than medium-sized coal dust samples were collected from mines in 7 of the 10 MSHA Districts and represented approximately 26% of the overall mines sampled (16 of the 61 mines).

The results of the LLEM large-scale explosion tests including the medium-sized coal dust are shown in Figure 11. Given the experimental test conditions, the curve is the boundary between mixtures that did propagate an explosion (below line) and mixtures that did not propagate an explosion (above line). The coal dust particle size has a substantial impact on the propagation potential for coal dust. As the coal dust particle size decreases, increasing amounts of rock dust are necessary to render the coal/rock dust mixture inert. The greatest impact is evident between the particle size of the coarse (20% minus 200 mesh or 75 μm) coal dust and the pulverized (80% minus 200 mesh or 75 μm) coal dust. To ensure non-propagation within the LLEM, the coarse coal dust required at least 70% TIC and the pulverized coal dust required greater than a 79% TIC and less than a 81.5% TIC. Once the 80% minus 200 mesh benchmark had been reached, no additional TIC was required to prevent flame propagation with further decrease in coal dust particle size. One can clearly see when comparing Figure 1 with the earlier BEM data to Figure 11 of the recent LLEM data that the TIC increases from about 60% to 70% TIC at the coarse coal particle size end of the figures, while TIC remains at about 80% at the fine coal particle size end of both figures.

The 80% limit is also consistent with explosion temperature thermodynamic limit models for coal and rock dust put forward by Richmond et al. [1975; 1979], Hertzberg et al. [1988], Conti et al. [1991], and Sapko et al. [2000]. The models were essentially based on a thermal balance between the heat generated during the combustion of coal dust and heat absorbed by the incombustible material.

LLEM inerting studies using a medium-sized coal dust showed that at least 76.4% TIC (Table C-4) is required to prevent explosion propagation. If one considers the finest size intake air way dust collected during the recent survey (63% minus 200 mesh from Table A-2), data in Figure 11 indicates that approximately 80% TIC would be required to prevent explosion propagation.

Summary

Dust explosibility is strongly dependent on the size distribution of the coal particles in a coal and rock dust mixture. Underground coal mining technology has changed since the 1920s; that is, coal mining has become highly mechanized, creating coal dust with more small size fractions than those of the 1920s. Despite this change in technology, particle size surveys from the early 1900s are still being used as the basis for current rock dusting regulations. Although total incombustible content is an important determinant of explosion propagation, coal dust particle size also needs to be considered as an essential part of an explosibility assessment in underground coal mines. The present coal size study indicates that the coal dust in intake airways of U.S. mines is finer than that measured by Rice and Greenwald [1929] in the 1920s. Moreover, particle size distributions can vary with coal seam and rank, as shown in Table 2. Current rock dust regulations mandating a 65% TIC dust mixture do not fully protect miners since LLEM tests have shown that even a ~68% TIC dust mixture with coarse Pittsburgh seam coal dust (20% minus 200 mesh) will propagate dust explosions. LLEM inerting experiments also demonstrated that at least 76.4% TIC is required to prevent explosion propagation for medium-sized coal dust (38% minus 200 mesh)—that is, an average of the finer dust found in modern intake areas. For return airways, the current requirement of at least 80% TIC is still sufficient in the absence of methane.

LLEM experiments for high volatile coals have also shown that the TIC required to prevent flame propagation becomes much less dependent on coal particle size as the TIC approaches and exceeds 80%. Therefore, experimental results support at least an 80% TIC requirement for both intake and return airways in the absence of methane.

Recommendation

Large-scale explosion testing in the Bruceton and Lake Lynn Experimental Mines confirm intake airways require more incombustible content to render the coal dust inert than the 65% TIC specified in current regulations.

NIOSH recommends an 80% TIC in intake airways based on:
- Explosion temperature thermodynamic limit models for coal and rock dust mixtures,
- Extensive in-mine coal dust particle size surveys, and
- Multiple explosion experiments at the Lake Lynn Laboratory.

Acknowledgments

The authors acknowledge the invaluable technical guidance from Harry Verakis, General Engineer, of the MSHA Approval and Certification Center, Triadelphia, WV, and Terry Montgomery, Supervisor Chemist, of the MSHA analytical laboratory, Mt. Hope, WV, for providing the dust samples and the incombustible percentage data. We acknowledge the many MSHA inspectors for collecting the dust samples in coal mines throughout the United States. The authors thank E. Cathy Hedrick, Information Technology Specialist, of MSHA for providing the intake and return sample data over the past 3 years. The authors also acknowledge the laboratory assistance of the following NIOSH-OMSHR personnel for their efforts in the acid-leaching and size analysis of the dust samples: Kenneth Helfrich, Physical Science Technician, and Electronic Technicians Jon Jobko and Andrew Mazzella.

The authors also thank the following NIOSH-OMSHR personnel whose contributions made the 2008 LLEM inerting tests possible: William Slivensky, Frank Karnack, Donald Sellers, and James Addis, Physical Science Technicians at LLL, for their participation in the installation of sensors, the construction of the methane ignition zones, the installation of the gas sampling lines, the dispersal of the dust mixtures for each test, performing experiments, and post-explosion dust sampling; Kenneth Jackson, Electronics Technician at LLL, for sensor calibrations, operation of the data acquisition systems, and data analyses; and Paul Stefko, Foreman I – Miner Mechanic, of the Safety Research Coal Mine (SRCM) at OMSHR; Jack Teatino and Joseph Sabo, SRCM Miner Mechanics, for mining and then grinding and/or pulverizing the coal to the proper particle size and preparing the coal/rock dust mixtures.

The authors acknowledge the following mechanical technicians with Ki Corporation (a NIOSH contractor) for their thorough cleanup of the debris and the wash down of the entry following each large-scale explosion test: Timothy Glad (lead), James Rabon, Bernard Lambie, and Brandin Lambie.

References

ASTM International [2008]. Standard classification of coals by rank, D388–05. In: Annual Book of ASTM Standards, vol. 5.06. West Conshohocken, PA: ASTM International.

Cashdollar KL [1996]. Coal dust explosibility. J Loss Prev Process Ind 9(1):65–76.

Cashdollar KL, Chatrathi K [1993]. Minimum explosible dust concentrations measured in 20-L and 1-m^3 chambers. Combustion Science and Technology, 87(1–6):157–171.

Cashdollar KL, Hertzberg M [1989]. Laboratory study of rock dust inerting requirements: effects of coal volatility, particle size, and methane addition. In: Proceedings of the 23rd International Conference of Safety in Mines Research Institutes (Washington, DC, September 11–15, 1989). Pittsburgh, PA: U.S. Department of the Interior, Bureau of Mines, pp. 965–977.

Cashdollar KL, Hertzberg M, Zlochower IA [1988]. Effect of volatility on dust flammability limits for coals, gilsonite, and polyethylene. In: Proceedings of the 22nd International Symposium on Combustion. The Combustion Institute, Pittsburgh, PA, 1988, pp. 1757–1765.

Cashdollar KL, Sapko MJ, Weiss ES, Hertzberg M [1987]. Laboratory and mine dust explosion research at the Bureau of Mines. In: Cashdollar KL, Hertzberg M, eds. Industrial dust explosions. West Conshohocken, PA: American Society for Testing and Materials (ASTM), Special Technical Publication (STP) 958, pp. 107–123.

Cashdollar KL, Weiss ES, Greninger NB, Chatrathi K [1992a]. Laboratory and large-scale dust explosion research. Plant/Operations Progress, 11, pp. 247–255.

Cashdollar KL, Weiss ES, Greninger NB, Chatrathi K [1992b]. Laboratory and large-scale dust explosion research. In: Proceedings of the 26th Annual Loss Prevention Symposium (New Orleans, LA). Paper No. 72C. AIChE Spring National Meeting, 22 pp.

Cashdollar KL, Weiss ES, Greninger NB, Hertzberg M, Sapko MJ [1992c]. Laboratory and large-scale explosion research at the U.S. Bureau of Mines. In: Proceedings of The First World Seminar on the Explosion Phenomenon and on the Application of Explosion Protection Techniques in Practice (Brussels, Belgium). Section 1, 16 pp.

Cashdollar KL, Weiss ES, Montgomery TG, Going JE [2007]. Post-explosion observations of experimental mine and laboratory coal dust explosions. J Loss Prev Process Ind 20(1):607–615.

CFR, Code of Federal Regulations. Washington, DC: U.S. Government Printing Office, Office of the Federal Register.

Conti RS, Zlochower IA, Sapko MJ [1991]. Rapid sampling of products during coal mine explosions. Combustion Science and Technology, 75(4–6):195–209.

Cybulski WG [1975]. Coal dust explosions and their suppression. Translated from Polish. Warsaw, Poland: National Center for Scientific, Technical and Economic Information. NTIS No. TT 73–54001.

Dobroski H Jr., Stephan CR, Conti RS [1996]. Historical summary of coal mine explosions in the United States. 1981–1994. Pittsburgh, PA: U.S. Department of the Interior, Bureau of Mines, IC 9440.

Federal Coal Mine Safety Act of 1952 (Public Law 82–552).

Greninger NB, Cashdollar KL, Weiss ES, Sapko MJ [1991]. Suppression of dust explosions involving fuels of intermediate and high volatile content. In: Proceedings of the Fourth International Colloquium on Dust Explosions, Polish Academy of Sciences, Porabka-Kozubnik, Poland, November 4–9, 1990, pp. 208–228.

Hertzberg M, Cashdollar KL [1987]. Introduction to dust explosions. In: Cashdollar KL, Hertzberg M, eds. Industrial dust explosions. West Conshohocken, PA: American Society for Testing and Materials (ASTM), Special Technical Publication (STP) 958, pp. 5–32.

Hertzberg M, Zlochower IA, Cashdollar KL [1988a]. Volatility model for coal dust flame propagation and extinguishment. In: Proceedings of the 21st International Symposium on Combustion (University of Munich, Germany). Pittsburgh, PA: The Combustion Institute, pp. 325–353.

Hertzberg M, Zlochower IA, Edwards JC [1988b]. Coal pyrolysis mechanisms and temperatures. Pittsburgh, PA: U.S. Department of the Interior, Bureau of Mines, RI 9169.

Lebecki K [1991]. Gas dynamics of coal dust explosion—theory and experiment. In: Proceedings of the 24th International Conference of Safety in Mines Research Institutes (Donetsk, USSR, September 23–28), vol. 1, pp. 357–373.

Liebman I, Richmond JK [1974]. Suppression of coal dust explosions by passive water barriers in a single entry mine. Pittsburgh, PA: U.S. Department of the Interior, Bureau of Mines, RI 7815.

Light TE, Herndon RC, Guley AR Jr., Cook GL, Odum MA, Bates RM Jr., Schroeder ME, Campbell CD, Pruitt ME [2007]. Report of investigation, fatal underground coal mine explosions, May 20, 2006. Darby mine No. 1, Kentucky Darby LLC, Holmes Mill, Harlan County, Kentucky, ID No. 15–18185. Arlington, VA: U.S. Department of Labor, Mine Safety and Health Administration.

McKinney R, Crocco W, Stricklin KG, Murray KA, Blankenship ST, Davidson RD, Urosek JE, Stephan CR, Beiter DA [2002]. Report of investigation, fatal underground coal mine explosions, September 23, 2001. No. 5 Mine, Jim Walters Resources, Inc., Brookwood, Tuscaloosa County, Alabama, ID No. 01–01322. Arlington, VA: U.S. Department of Labor, Mine Safety and Health Administration.

Michelis J [1996]. Large scale experiments with coal dust explosions in connection with road-T-junctions. In: The Seventh International Colloquium on Dust Explosions; part of the International Symposium on Hazards, Prevention and Mitigation of Industrial Explosions (Bergen, Norway, June 23–26, 1996). Christian Michelsen Research, pp. 8.50–8.59.

Michelis J, Margenburg B, Müller G, Kleine W [1987]. Investigations into the buildup and development conditions of coal dust explosions in a 700-m underground gallery. In: Cashdollar KL, Hertzberg M, eds. Industrial dust explosions. West Conshohocken, PA: American Society for Testing and Materials (ASTM), Special Technical Publication (STP) 958, pp. 124–137.

Mine Safety Board [1937]. Recommendations of the United States Bureau of Mines on certain questions of safety as of Oct. 1, 1936. Pittsburgh, PA: U.S. Department of the Interior, Bureau of Mines, IC 6946.

Montgomery T [2005]. Personal communication on incombustible analysis procedures.

MSHA [2008] Handbook Series, Handbook Number PH–08–V–1, General Coal Mine Inspection Procedures and Inspection Tracking System, January 1, 2008, pp. 45–66. http://www.msha.gov/readroom/handbook/handbook.htm

Nagy J [1981]. The explosion hazard in mining. Pittsburgh, PA: U.S. Department of Labor, Mine Safety and Health Administration, IR 1119.

Reed D, Michelis J [1989]. Comparative investigation into explosibility of brown coal and bituminous coal dust in surface and underground test installations. In: Proceedings of the 23rd International Conference of Safety in Mines Research Institutes (Washington, DC, September 11–15, 1989). Pittsburgh, PA: U.S. Department of the Interior, Bureau of Mines, pp. 941–964.

Rice GS, Greenwald HP [1929]. Coal-dust explosibility factors indicated by experimental mine investigations 1911 to 1929. Pittsburgh, PA: U.S. Department of the Interior, Bureau of Mines, Technical Paper 464.

Rice GS, Jones LM, Egy WL, Greenwald HP [1922]. Coal-dust explosion tests in the experimental mine 1913 to 1918, inclusive. Pittsburgh, PA: U.S. Department of the Interior, Bureau of Mines, Bulletin 167.

Rice GS [1927]. Effective rock-dusting of coal mines. Pittsburgh, PA: U.S. Department of the Interior, Bureau of Mines, IC 6039.

Richmond JK, Liebman I, Miller LF [1975]. Effect of rock dust on the explosibility of coal dust. Pittsburgh, PA: U.S. Department of the Interior, Bureau of Mines, RI 8077, 34 pp.

Richmond JK, Liebman I, Bruszak AE, Miller LF [1979]. A physical description of coal mine explosions, part II. Seventeenth symposium (international) on combustion. Pittsburgh, PA: The Combustion Institute, pp. 1257–1268.

Sapko MJ, Weiss ES, Watson RW [1987a]. Explosibility of float coal dust distributed over a coal-rock dust substratum. In: Proceedings of the 22nd International Conference of Safety in Mines Research Institutes (Beijing, China, November 2–6, 1987). Beijing, China: China Coal Industry Publishing House, pp. 459–468.

Sapko MJ, Weiss ES, Watson RW [1987b]. Size scaling of gas explosions: Bruceton Experimental Mine versus the Lake Lynn Mine. Pittsburgh, PA: U.S. Department of the Interior, Bureau of Mines, RI 9136. NTIS No. PB 88–230248.

Sapko MJ, Greninger NB, Watson RW [1989]. Review paper: Prevention and Suppression of Coal Mine Explosions. In: Proceedings of the 23rd International Conference of Safety in Mines Research Institutes (Washington, DC, September 11–15, 1989). Pittsburgh, PA: U.S. Department of the Interior, Bureau of Mines, pp. 791–807.

Sapko MJ, Weiss ES, Cashdollar KL, Zlochower IA [1998]. Experimental mine and laboratory dust explosion research at NIOSH. In: Proceedings of the International Symposium on Hazards, Prevention, and Mitigation of Industrial Explosions: Eighth International Colloquium on Dust Explosions (Schaumburg, IL, September 21–25, 1998). Safety Consulting Engineers, pp. 120–142.

Sapko MJ, Cashdollar KL, Green GM [2007]. Coal dust particle size survey of U.S. mines. J Loss Prev Process Ind, *20*(1): 616–620.

Sapko M J, Weiss ES, Cashdollar KL, Zlochower IA [2000]. Experimental mine and laboratory dust explosion research at NIOSH. J Loss Prev Process Ind. *13*(1): 229–242.

Triebsch G, Sapko MJ [1990]. Lake Lynn Laboratory: a state-of-the-art mining research laboratory. In: Proceedings of the International Symposium on Unique Underground Structures, vol. 2, Golden, CO, June 11–15, 1990. Colorado School of Mines, pp. 75–1 to 75–21.

U.S. Code of Federal Regulations [2010]. Title 30 CFR, Part 75, Section 75.400 and 75.403.

U.S. Code of Federal Regulations [2010]. Title 30 CFR, Part 75, Section 75.2.

U.S. Congress [1969]. Federal Coal Mine Health and Safety Act of 1969. Public Law 91–173, 83 Stat. 742, pp. 32–33.

U.S. Congress [1977], Coal Mine Safety and Health Act of 1977, Public Law 95–164, Section 304, p. 49.

Weiss ES, Greninger NB, Sapko MJ [1989]. Recent results of dust explosion studies at the Lake Lynn Experimental Mine. In: Proceedings of the 23rd International Conference of Safety in Mines Research Institutes (Washington, DC, September 11–15, 1989). Pittsburgh, PA: U.S. Department of the Interior, Bureau of Mines, pp. 843–856.

Appendix A:
Analyses of Size of Coal Dust Particles from Mine Intake Airways

Table A-1. Analyses of size of coal dust particles from intake airways in six MSHA District 2 mines

Mine	Production, Mt/yr	Incombustible, %	Soluble, %	Size analysis −270 mesh or < 53 µm, %	Size analysis −70 mesh or < 212 µm, %	Ash, %	Corrected size analysis −270 mesh or < 53 µm, %	−200 mesh or < 75 µm, %	−140 mesh or < 106 µm, %	−100 mesh or < 150 µm, %	−70 mesh or < 212 µm, %	−50 mesh or < 300 µm, %	−40 mesh or < 425 µm, %	−30 mesh or < 600 µm, %	D_{med}, µm
A	>1	74	55	35	64	40	21	25	33	44	56	69	83	92	178
		81	73	43	70	42	27	34	44	53	64	76	89	97	136
B	>1	54	52	38	71	22	25	32	43	55	68	80	91	97	130
		60	42	31	61	27	19	25	35	45	57	70	84	95	173
		82	69	41	72	40	25	31	41	52	64	76	89	97	143
		56	24	37	60	37	25	29	36	45	55	67	82	94	180
		72	61	33	59	30	22	27	34	44	54	67	82	94	186
		85	40	48	79	73	22	31	40	50	60	77	89	97	151
C	>1	86	58	31	62	23	21	26	35	44	57	72	85	95	176
		88	79	46	74	46	26	32	43	55	67	77	87	95	130
		75	48	35	62	47	23	27	35	44	57	70	86	97	177
		64	30	27	50	49	18	21	27	35	46	60	79	94	237
D	>1	70	51	39	65	36	23	30	36	44	55	68	82	93	184
		93	85	35	71	39	23	28	38	50	63	73	84	94	150
		67	42	34	59	38	24	28	34	44	55	70	85	96	182
		50	24	46	71	30	33	38	44	54	66	78	90	98	130
E	<1	75	66	38	64	21	30	36	43	53	63	73	84	95	135
F	<1	90	57	30	65	70	17	24	33	43	55	69	80	90	186
		90	73	28	61	58	15	22	31	41	54	70	86	96	191
		88	69	39	66	58	23	31	38	48	58	71	85	96	159
average for MSHA District 2							23	29	37	47	59	72	85	95	165
standard deviation							4	4	5	5	5	5	3	2	27

Notes:
The incombustible content is the value measured by the MSHA Mt. Hope laboratory.
The soluble content is the percentage that is soluble in hydrochloric acid (i.e., the calcium carbonate content of the limestone or marble rock dust), as measured at OMSHR.
The ash includes the ash in the coal plus the insoluble mineral material, as measured at OMSHR.

Table A-2. Analyses of size of coal dust particles from intake airways in seven MSHA District 3 mines

Mine	Production, Mt/yr	Incombustible, %	Soluble, %	Size analysis			ash, %	Corrected size analysis							D_{med}, μm
				−270 mesh or < 53 μm, %	−70 mesh or < 212 μm, %			−270 mesh or < 53 μm, %	−200 mesh or < 75 μm, %	−140 mesh or < 106 μm, %	−100 mesh or < 150 μm, %	−70 mesh or < 212 μm, %	−50 mesh or < 300 μm, %	−40 mesh or < 425 μm, %	−30 mesh or < 600 μm, %	
A	>1	55	27	29	61		21	20	27	35	46	60	75	90	98	165
B	>1	68	47	41	68		37	26	30	37	47	59	74	88	97	164
		70	44	31	61		46	17	22	30	40	53	70	86	97	199
C	>1	82	57	39	66		56	24	31	38	47	57	71	86	96	169
		97	96	42	71		35	28	38	48	58	70	75	85	93	113
		95	95	67	84		25	52	63	71	77	81	86	93	98	50
		90	81	47	77		44	28	36	45	57	66	78	89	97	123
		87	73	51	77		55	26	32	39	49	56	69	82	94	160
		86	72	37	71		47	23	32	41	52	63	78	91	98	141
		88	77	45	74		47	32	40	46	54	63	77	89	97	125
D	>1	83	81	45	76		20	32	39	49	61	72	82	91	97	108
		77	68	52	82		23	38	45	55	66	78	88	94	98	89
		91	74	40	72		55	20	25	35	47	59	73	87	95	164
		72	55	42	67		30	27	32	39	49	60	72	86	96	156
		46	11	37	62		33	24	29	36	45	56	69	84	96	175
		41	10	34	62		32	23	28	36	46	57	71	87	97	171
E	>1	80	59	46	82		44	27	34	46	60	74	86	95	99	117
		79	63	32	66		33	20	26	35	47	62	75	86	93	161
F	>1	83	75	43	75		40	25	32	43	54	66	78	89	97	134
		75	67	43	69		50	25	31	41	49	60	70	83	94	155
G	>1	58	39	29	55		23	21	28	34	43	54	65	79	92	189
		72	63	20	44		19	13	18	24	32	42	56	75	92	259
					average for MSHA District 3				26	33	41	51	62	74	87	96	149
					standard deviation				8	9	10	10	9	7	5	2	42

Notes: The incombustible content is the value measured by the MSHA Mt. Hope laboratory.
The soluble content is the percentage that is soluble in hydrochloric acid (i.e., the calcium carbonate content of the limestone or marble rock dust), as measured at OMSHR.
The ash includes the ash in the coal plus the insoluble mineral material, as measured at OMSHR.

Table A-3. Analyses of size of coal dust particles from intake airways in seven MSHA District 4 mines

Mine	Production, Mt/yr	Incombustible, %	Soluble, %	Size analysis		ash, %	Corrected size analysis							D_{med}, μm
				-270 mesh or < 53 μm, %	-70 mesh or < 212 μm, %		-270 mesh or < 53 μm, %	-200 mesh or < 75 μm, %	-140 mesh or < 106 μm, %	-100 mesh or < 150 μm, %	-70 mesh or < 212 μm, %	-50 mesh or < 300 μm, %	-40 mesh or < 425 μm, %	-30 mesh or < 600 μm, %	
A	>1	65	51	42	70	30	26	34	44	54	64	75	86	96	131
		68	54	36	59	26	22	28	35	43	52	65	80	94	195
		64	35	45	69	43	29	36	44	54	64	75	87	97	133
		70	61	33	63	25	22	29	37	48	61	73	87	97	160
		89	83	52	80	35	34	41	50	62	74	85	95	99	106
		85	81	43	68	25	29	35	42	52	64	77	90	98	138
B	<1	82	40	40	63	61	24	30	36	45	55	68	82	94	181
		75	—	37	65	65	23	28	36	45	56	70	85	96	176
C	<1	68	18	30	54	59	21	25	32	41	52	65	82	96	199
		70	15	38	59	61	25	28	35	43	53	66	81	95	192
		69	21	32	56	57	18	22	27	36	47	61	79	94	231
		81	35	34	60	70	20	25	33	42	53	69	85	97	196
		78	33	38	60	68	26	29	35	44	56	67	83	96	179
D	<1	79	38	38	60	63	24	28	34	44	54	66	82	95	186
E	<1	47	9	47	82	37	29	36	47	61	75	88	95	99	114
		57	30	25	54	29	16	21	28	37	48	63	79	94	224
		37	3	32	59	27	20	25	33	43	54	67	81	94	188
		53	18	24	59	40	12	16	23	33	48	68	86	97	221
F	>1	77	58	45	71	48	31	36	44	55	65	78	91	98	129
		87	73	44	70	52	29	34	43	53	63	77	91	98	137
G	<1	40	23	45	84	23	35	43	53	67	83	93	97	99	98
		37	16	43	75	22	29	36	45	57	71	85	95	99	123
		49	—	33	69	29	24	30	39	51	67	83	93	98	147
						average for MSHA District 4	25	30	38	48	60	73	87	97	165
						standard deviation	6	6	7	9	10	9	6	2	39

Notes: The incombustible content is the value measured by the MSHA Mt. Hope laboratory.
The soluble content is the percentage that is soluble in hydrochloric acid (i.e., the calcium carbonate content of the limestone or marble rock dust), as measured at OMSHR.
The ash includes the ash in the coal plus the insoluble mineral material, as measured at OMSHR.

Table A-4. Analyses of size of coal dust particles from intake airways in six MSHA District 5 mines

Mine	Production, Mt/yr	Incombustible, %	Soluble, %	Size analysis			Corrected size analysis								D_{med}, µm
				−270 mesh or < 53 µm, %	−70 mesh or < 212 µm, %	ash, %	−270 mesh or < 53 µm, %	−200 mesh or < 75 µm, %	−140 mesh or < 106 µm, %	−100 mesh or < 150 µm, %	−70 mesh or < 212 µm, %	−50 mesh or < 300 µm, %	−40 mesh or < 425 µm, %	−30 mesh or < 600 µm, %	
A	<1	64	43	32	58	34	25	31	39	49	60	72	84	95	156
		57	28	26	52	44	23	29	36	45	57	68	82	94	173
		68	60	38	61	32	28	34	41	50	59	71	85	96	150
B	>1	66	57	47	72	30	27	30	39	49	61	73	86	96	154
		63	63	52	78	20	40	47	55	65	75	83	92	98	87
		87	85	31	56	39	18	26	33	42	51	64	81	94	203
		48	43	37	62	26	24	31	39	50	60	72	85	95	151
		35	24	23	48	26	16	21	27	37	47	61	78	92	227
		65	56	36	62	32	27	33	40	51	62	75	88	97	145
C	<1	83	73	60	94	29	40	54	70	86	93	96	98	99	68
		69	54	38	77	28	26	32	42	56	75	88	95	98	132
		78	61	44	83	32	30	36	48	64	80	90	95	98	110
D	<1	52	29	33	59	31	21	28	35	45	56	69	83	95	176
		57	28	34	62	34	21	27	34	45	57	70	84	96	175
E	>1	82	76	35	62	27	23	29	36	46	57	68	81	93	173
		72	58	32	65	21	22	28	36	46	62	76	89	98	164
		67	52	26	62	25	21	22	36	45	58	70	84	96	172
F	<1	77	58	38	61	43	27	32	38	47	58	69	84	96	166
		80	67	35	60	43	25	30	37	46	57	69	83	96	170
		76	63	30	58	33	16	26	34	44	55	67	81	94	183
average for MSHA District 5							25	31	40	50	62	74	86	96	157
standard deviation							6	8	9	11	11	9	5	2	36

Notes: The incombustible content is the value measured by the MSHA Mt. Hope laboratory.
The soluble content is the percentage that is soluble in hydrochloric acid (i.e., the calcium carbonate content of the limestone or marble rock dust), as measured at OMSHR.
The ash includes the ash in the coal plus the insoluble mineral material, as measured at OMSHR.

Table A-5. Analyses of size of coal dust particles from intake airways in five MSHA District 6 mines

Mine	Production, Mt/yr	Incombustible, %	Soluble, %	Size analysis			Corrected size analysis							D_{med}, μm
				−270 mesh or < 53 μm, %	−70 mesh or < 212 μm, %	ash, %	−270 mesh or < 53 μm, %	−200 mesh or < 75 μm, %	−140 mesh or < 106 μm, %	−100 mesh or < 150 μm, %	−70 mesh or < 212 μm, %	−50 mesh or < 300 μm, %	−40 mesh or < 425 μm, %	−30 mesh or < 600 μm, %	
A	>1	63	31	37	64	41	20	25	33	42	53	66	83	96	195
		54	26	38	61	35	23	28	35	43	54	66	81	95	188
		51	14	33	58	45	21	26	33	42	53	66	82	95	193
B	<1	36	17	28	56	18	19	25	31	41	52	65	82	96	200
		40	22	28	56	20	18	23	29	37	50	65	82	96	214
		37	20	38	63	17	27	33	40	48	58	69	82	94	164
		35	21	40	64	17	29	35	42	50	60	71	85	96	150
		35	19	35	62	18	23	29	36	46	57	70	83	94	173
		37	20	38	64	19	27	32	39	48	58	72	87	97	160
		73	60	42	70	27	27	33	41	51	62	75	87	96	145
C	>1	77	50	30	55	56	17	22	28	37	48	64	81	95	220
		73	20	42	65	65	28	32	39	50	61	73	86	97	150
		73	24	35	59	64	18	24	30	39	49	64	81	96	215
D	>1	76	25	46	82	67	25	29	38	51	63	77	89	97	145
		76	29	47	72	69	27	34	45	56	64	73	85	97	124
		74	21	42	75	67	22	30	38	48	61	78	89	97	161
		71	17	50	77	65	28	33	41	52	62	74	84	95	142
		72	12	52	83	67	29	34	45	59	69	79	87	96	120
E	<1	84	81	60	79	30	45	55	63	70	77	83	92	98	64
		84	75	50	79	34	34	42	51	62	71	81	90	97	102
		64	47	40	64	26	29	34	41	50	59	70	84	96	151
		86	77	36	69	42	23	29	37	49	61	75	88	97	155
		56	41	37	67	21	26	32	39	50	62	77	90	98	150
		56	41	36	62	23	26	32	39	48	59	71	85	96	162
average for MSHA District 6							25	31	39	49	59	72	85	96	160
standard deviation							6	7	8	8	7	6	3	1	37

Notes: The incombustible content is the value measured by the MSHA Mt. Hope laboratory.
The soluble content is the percentage that is soluble in hydrochloric acid (i.e., the calcium carbonate content of the limestone or marble rock dust), as measured at OMSHR.
The ash includes the ash in the coal plus the insoluble mineral material, as measured at OMSHR.

Table A-6. Analyses of size of coal dust particles from intake airways in five MSHA District 7 mines

Mine	Production, Mt/yr	Incombustible, %	Soluble, %	Size analysis			Corrected size analysis							
				-270 mesh or < 53 µm, %	-70 mesh or < 212 µm, %	ash, %	-270 mesh or < 53 µm, %	-200 mesh or < 75 µm, %	-140 mesh or < 106 µm, %	-100 mesh or < 150 µm, %	-70 mesh or < 212 µm, %	-50 mesh or < 300 µm, %	-40 mesh or < 425 µm, %	-30 mesh or < 600 µm, %	D_{med}, µm
A	<1	79	65	41	67	51	23	29	37	46	55	68	82	95	175
		79	65	44	74	50	24	29	39	51	60	74	87	96	147
		81	62	41	68	52	22	29	38	48	58	72	85	96	164
		78	60	37	65	46	21	27	35	45	55	69	84	95	179
B	<1	92	80	62	78	63	52	56	63	69	75	80	84	90	46
		92	83	66	82	63	49	54	62	69	76	82	88	94	59
		92	78	63	83	62	48	54	61	69	78	82	88	94	60
C	<1	89	62	44	77	65	20	27	37	50	61	74	87	97	149
		87	66	55	83	59	29	37	47	59	71	84	93	98	117
		96	87	45	74	74	24	28	37	52	63	72	85	95	143
		90	78	59	86	61	29	36	43	58	69	81	90	97	124
		91	78	45	70	59	22	29	42	51	59	71	85	96	144
D	<1	61	24	39	64	49	22	27	35	45	55	69	84	95	179
		74	28	36	63	60	20	26	33	42	53	67	82	95	195
		77	38	38	56	58	19	25	32	40	52	66	82	95	200
E	<1	88	69	59	85	59	31	35	47	58	67	78	88	96	117
		91	74	64	83	62	36	41	49	58	66	76	87	96	110
		82	70	34	61	39	19	23	29	39	49	65	81	94	215
		84	64	57	75	53	33	37	44	53	62	74	86	96	136
average for MSHA District 7							29	34	43	53	62	74	86	95	140
standard deviation							10	10	10	9	8	6	3	2	48

Notes: The incombustible content is the value measured by the MSHA Mt. Hope laboratory.
The soluble content is the percentage that is soluble in hydrochloric acid (i.e., the calcium carbonate content of the limestone or marble rock dust), as measured at OMSHR.
The ash includes the ash in the coal plus the insoluble mineral material, as measured at OMSHR.

Table A-7. Analyses of size of coal dust particles from intake airways in six MSHA District 8 mines

Mine	Production, Mt/yr	Incombustible, %	Soluble, %	Size analysis		ash, %	Corrected size analysis							D_{med}, µm
				−270 mesh or < 53 µm, %	−70 mesh or < 212 µm, %		−270 mesh or < 53 µm, %	−200 mesh or < 75 µm, %	−140 mesh or < 106 µm, %	−100 mesh or < 150 µm, %	−70 mesh or < 212 µm, %	−50 mesh or < 300 µm, %	−40 mesh or < 425 µm, %	−30 mesh or < 600 µm, %	
A	>1	92	81	27	54	64	14	18	25	35	47	59	76	92	234
		97	93	38	66	52	21	27	37	48	60	70	83	95	160
B	>1	82	49	49	69	51	23	27	34	42	51	64	80	94	208
		81	49	53	77	57	24	29	39	48	58	70	83	95	161
C	>1	75	45	42	71	46	24	27	35	45	58	73	86	97	171
		68	30	43	79	56	22	27	38	51	65	78	91	98	145
D	>1	67	32	50	68	46	29	33	39	47	57	68	84	96	167
		78	57	53	74	46	33	38	45	54	65	77	90	98	130
		65	21	41	63	51	20	26	33	41	52	68	84	96	198
		68	33	46	69	45	25	30	39	48	58	70	84	95	162
E	>1	84	19	47	73	77	26	30	37	49	59	73	89	98	156
		82	23	41	66	75	25	30	36	46	56	71	86	97	175
		76	22	43	67	66	25	29	36	46	57	69	85	96	172
		79	23	49	70	68	27	30	37	46	55	68	83	96	178
F	>1	86	63	55	80	63	30	33	43	54	65	78	89	96	132
		73	43	50	74	43	27	32	41	50	60	72	85	95	149
		88	63	54	83	57	24	30	44	56	67	81	92	98	127
		67	25	36	61	44	15	17	24	34	47	62	80	95	230
average for MSHA District 8							24	29	37	47	57	71	85	96	170
standard deviation							5	5	5	6	6	6	4	2	31

Notes: The incombustible content is the value measured by the MSHA Mt. Hope laboratory.
The soluble content is the percentage that is soluble in hydrochloric acid (i.e., the calcium carbonate content of the limestone or marble rock dust), as measured at OMSHR.
The ash includes the ash in the coal plus the insoluble mineral material, as measured at OMSHR.

Table A-8. Analyses of size of coal dust particles from intake airways in seven MSHA District 9 mines

Mine	Production, Mt/yr	Incombustible, %	Soluble, %	Size analysis			Corrected size analysis							D_{med}, μm
				−270 mesh or < 53 μm, %	−70 mesh or < 212 μm, %	ash, %	−270 mesh or < 53 μm, %	−200 mesh or < 75 μm, %	−140 mesh or < 106 μm, %	−100 mesh or < 150 μm, %	−70 mesh or < 212 μm, %	−50 mesh or < 300 μm, %	−40 mesh or < 425 μm, %	−30 mesh or < 600 μm, %	
A	>1	84	74	42	71	48	21	28	36	47	56	69	83	94	170
		59	27	35	62	40	19	26	34	44	54	67	82	94	187
		88	77	39	71	49	23	31	40	51	61	75	88	97	147
		81	65	44	70	44	27	33	42	53	61	74	87	96	135
B	>1	83	70	32	62	26	19	26	34	45	56	70	85	97	176
C	>1	92	85	46	74	45	23	30	42	52	61	74	87	97	139
		60	53	35	63	25	20	26	34	44	57	71	87	97	178
		71	53	45	71	38	25	31	41	51	63	75	89	97	146
		53	25	42	66	33	24	28	36	46	56	70	84	95	173
		85	81	40	68	39	21	27	39	49	59	72	86	96	153
D	>1	81	87	34	63	36	16	22	30	40	49	62	77	92	220
		78	72	37	64	22	23	30	37	47	58	70	84	95	166
		78	72	38	70	21	25	31	41	52	64	78	90	98	141
		82	77	33	61	23	20	26	33	43	54	66	82	95	190
E	>1	76	68	35	61	15	25	30	37	46	57	68	83	94	172
		53	53	25	50	12	17	21	28	36	47	60	78	93	232
F	<1	40	31	26	55	9	17	24	31	41	53	68	84	95	196
		56	49	30	63	11	20	28	36	48	61	75	90	99	159
		47	34	29	59	10	20	27	34	44	56	70	86	97	179
G	<1	54	55	26	64	12	16	21	30	43	60	79	92	98	174
average for MSHA District 9							21	27	36	46	57	71	85	96	172
standard deviation							3	3	4	4	4	5	4	2	26

Notes: The incombustible content is the value measured by the MSHA Mt. Hope laboratory.
The soluble content is the percentage that is soluble in hydrochloric acid (i.e., the calcium carbonate content of the limestone or marble rock dust), as measured at OMSHR.
The ash includes the ash in the coal plus the insoluble mineral material, as measured at OMSHR.

Table A-9. Analyses of size of coal dust particles from intake airways in five MSHA District 10 mines

Mine	Production, Mt/yr	Incombustible, %	Soluble, %	Size analysis			Corrected size analysis								D_{med}, µm
				−270 mesh or < 53 µm, %	−70 mesh or < 212 µm, %	ash, %	−270 mesh or < 53 µm, %	−200 mesh or < 75 µm, %	−140 mesh or < 106 µm, %	−100 mesh or < 150 µm, %	−70 mesh or < 212 µm, %	−50 mesh or < 300 µm, %	−40 mesh or < 425 µm, %	−30 mesh or < 600 µm, %	
A	>1	67	49	42	73	41	25	30	39	50	62	77	90	98	148
		89	76	44	80	47	26	31	43	57	70	83	94	99	126
		72	na	42	73	41	23	29	39	50	59	71	84	95	147
		62	39	43	69	35	26	31	40	49	58	71	85	95	154
		74	52	41	72	44	23	27	37	48	58	74	88	96	161
		67	57	37	71	35	23	28	38	50	62	74	86	96	150
		89	82	52	82	41	29	41	56	66	75	85	94	99	93
		58	27	42	78	43	25	31	41	54	66	80	91	98	135
		56	32	50	81	40	14	22	37	52	65	79	90	97	144
		66	34	45	72	49	23	27	35	45	52	65	80	93	195
B	>1	86	74	37	71	41	21	27	36	48	60	73	85	94	159
		85	75	35	67	37	22	27	37	49	61	72	85	96	154
		84	72	37	71	35	25	30	41	51	62	75	87	96	145
		89	80	35	71	37	21	26	36	48	61	76	87	96	159
		87	78	36	69	29	22	27	37	48	60	73	86	96	159
		82	68	31	67	37	18	24	32	43	55	69	83	94	184
C	>1	75	52	35	63	38	21	26	33	43	53	66	81	94	196
		83	67	47	76	42	26	31	41	53	64	75	86	96	137
		89	77	54	80	45	27	38	52	60	67	77	87	96	100
		68	34	43	77	44	25	32	43	55	66	78	88	96	129
		79	47	33	70	56	17	24	32	43	53	68	81	94	195
		92	78	44	78	54	23	30	40	52	62	75	88	97	140
D	>1	86	74	36	67	42	22	28	38	49	59	71	83	92	156
		75	59	37	68	37	21	28	39	48	59	72	85	96	159
		86	70	34	70	42	19	25	38	49	59	72	87	96	153
E	>1	83	64	39	73	45	21	28	40	51	62	76	88	96	145
		75	51	48	71	51	27	31	38	47	57	69	82	94	168
		71	42	46	69	43	27	30	37	46	56	70	83	94	171
	average for MSHA District 10						23	29	39	50	61	74	86	96	152
	standard deviation						3	4	5	5	5	5	4	2	24

Notes: The incombustible content is the value measured by the MSHA Mt. Hope laboratory.
The soluble content is the percentage that is soluble in hydrochloric acid (i.e., the calcium carbonate content of the limestone or marble rock dust), as measured at OMSHR.
The ash includes the ash in the coal plus the insoluble mineral material, as measured at OMSHR.

Table A-10. Analyses of size of coal dust particles from intake airways in seven MSHA District 11 mines

Mine	Production, Mt/yr	Incombustible, %	Soluble, %	Size analysis			Corrected size analysis							D_{med}, μm
				−270 mesh or < 53 μm, %	−70 mesh or < 212 μm, %	ash, %	−270 mesh or < 53 μm, %	−200 mesh or < 75 μm, %	−140 mesh or < 106 μm, %	−100 mesh or < 150 μm, %	−70 mesh or < 212 μm, %	−50 mesh or < 300 μm, %	−40 mesh or < 425 μm, %	−30 mesh or < 600 μm, %	
A	>1	90	78	46	72	53	23	27	35	44	55	69	83	95	185
		91	79	47	79	57	25	31	41	52	66	80	91	98	141
B	>1	89	82	35	62	31	23	28	35	45	55	69	83	95	180
		54	30	31	57	32	21	28	35	44	55	67	81	94	184
C	>1	85	77	49	80	39	34	40	50	62	75	89	96	99	106
		86	80	51	83	33	35	42	52	65	78	90	96	99	99
D	>1	94	92	61	97	25	41	53	69	88	96	98	99	100	70
		58	41	40	71	28	29	36	45	55	68	84	95	99	128
E	>1	71	63	55	91	21	40	50	63	78	90	96	98	99	76
		89	78	42	73	25	29	36	45	56	70	84	94	99	126
		84	77	51	91	25	34	45	60	77	90	95	97	98	86
		91	84	43	66	38	26	31	38	47	57	68	81	93	168
		90	83	34	56	32	19	23	30	38	48	60	75	90	224
F	>1	68	63	47	93	10	33	45	60	79	92	98	99	100	85
		72	64	56	94	20	41	52	68	83	93	97	98	99	71
		62	55	47	85	16	35	45	59	74	86	93	97	99	85
		55	43	43	80	20	35	44	48	63	78	91	97	99	112
		66	64	28	70	7	19	27	36	50	69	90	98	99	149
		45	31	54	92	18	42	51	63	77	91	98	99	99	72
G	<1	47	28	35	64	22	24	29	37	46	58	75	90	98	169
		59	44	36	67	25	23	29	38	49	61	77	92	98	156
		58	38	50	87	29	33	42	55	71	84	94	98	99	94
		40	16	32	56	31	23	28	35	44	54	67	82	94	185
average for MSHA District 11							30	37	48	60	73	84	92	97	128
standard deviation							7	10	12	15	15	12	7	3	46

Notes: The incombustible content is the value measured by the MSHA Mt. Hope laboratory.
The soluble content is the percentage that is soluble in hydrochloric acid (i.e., the calcium carbonate content of the limestone or marble rock dust), as measured at OMSHR.
The ash includes the ash in the coal plus the insoluble mineral material, as measured at OMSHR.

Appendix B:
Analyses of Size of Coal Dust Particles from Mine Return Airways

Table B-1. Analyses of size of coal dust particles from return airways in 36 mines

Mine	Production, Mt/yr	Incombustible, %	Soluble, %	Size analysis				Corrected size analysis							D_{med}, μm
				−270 mesh or < 53 μm, %	−70 mesh or < 212 μm, %	ash, %	−270 mesh or < 53 μm, %	−200 mesh or < 75 μm, %	−140 mesh or < 106 μm, %	−100 mesh or < 150 μm, %	−70 mesh or < 212 μm, %	−50 mesh or < 300 μm, %	−40 mesh or < 425 μm, %	−30 mesh or < 600 μm, %	
1	>1	86	74	83	92	24	74	83	87	90	93	95	97	98	~30
		80	74	63	80	34	55	62	66	73	79	86	94	99	44
		87	76	72	88	41	62	69	74	79	83	88	93	98	42
2	>1	68	53	40	62	45	33	37	42	48	58	71	87	97	155
3	>1	63	40	57	72	35	44	47	52	59	66	75	85	95	91
		75	54	76	85	43	68	71	74	77	82	89	95	99	~20–25
4	<1	77	69	28	55	19	22	27	33	42	54	66	81	94	188
5	>1	82	79	85	93	12	78	84	88	90	93	95	98	99	~36
6	>1	80	75	59	76	15	52	58	63	68	75	82	90	97	49
7	>1	91	75	52	82	65	46	52	58	67	76	87	96	99	67
8	>1	72	45	63	78	43	52	57	62	68	75	82	91	98	45
9	>1	85	78	42	72	26	29	36	45	57	69	80	91	97	122
10	<1	46	14	38	83	32	24	30	40	55	79	95	99	100	135
11	>1	75	60	33	62	38	22	26	34	44	57	72	88	98	176
12	<1	37	24	27	54	18	19	24	32	42	53	66	80	93	193
13	<1	70	58	38	64	36	30	35	42	52	64	75	88	97	140
14	>1	71	75	42	68	22	30	35	43	53	63	75	87	96	137
15	>1	83	83	47	73	32	31	37	45	56	67	78	89	97	124
16	>1	76	54	41	64	49	24	27	34	42	54	68	84	96	190
17	>1	72	19	42	63	61	26	30	36	44	55	67	82	95	184
18	>1	50	29	32	56	27	24	30	37	45	55	67	82	95	178
19	<1	92	78	75	90	62	60	64	69	77	83	90	96	99	30
20	<1	89	68	36	62	60	20	25	33	43	54	68	85	97	189
21	<1	81	14	43	75	79	27	33	38	47	57	76	90	98	171
22	<1	86	62	56	75	65	32	35	43	52	61	72	84	94	141
23	>1	83	53	53	74	59	27	31	37	46	57	72	88	97	170
24	>1	64	40	38	63	132	22	29	36	45	56	69	83	95	178
25	>1	62	22	42	63	40	24	28	34	42	52	66	83	96	199
		56	22	43	65	37	25	29	35	44	55	69	87	97	182
26	>1	89	79	82	93	59	64	71	77	81	84	90	95	99	30
		70	70	47	69	28	33	40	47	56	64	75	87	95	121

Continued on Next Page

Mine	Production, Mt/yr	Incombustible, %	Soluble, %	Size analysis −270 mesh or < 53 μm, %	−70 mesh or < 2-2 μm, %	ash, %	Corrected size analysis −270 mesh or < 53 μm, %	−200 mesh or < 75 μm, %	−140 mesh or < 106 μm, %	−100 mesh or < 150 μm, %	−70 mesh or < 212 μm, %	−50 mesh or < 300 μm, %	−40 mesh or < 425 μm, %	−30 mesh or < 600 μm, %	D_{med}, μm
27	>1	79	69	46	73	29	31	37	45	56	67	80	91	98	124
28	>1	77	74	36	60	19	23	28	35	44	53	66	81	93	189
29	<1	66	53	30	54	12	21	26	32	40	50	63	80	94	211
30	<1	61	40	35	57	34	23	28	33	41	51	64	80	94	208
		62	16	39	66	54	25	30	37	46	57	73	88	97	171
31	>1	88	81	36	62	39	20	26	35	44	51	64	79	93	201
32	<1	65	21	44	70	52	25	29	35	44	54	69	83	95	186
33	<1	96	95	81	89	60	65	71	76	80	83	88	93	98	~25–30
34	>1	94	90	58	79	36	40	46	54	62	72	82	91	97	91
		93	88	50	79	26	33	40	49	61	74	87	95	99	109
		88	82	47	66	42	27	32	38	45	54	66	80	92	183
35	>1	39	25	38	66	17	28	33	41	50	62	77	91	98	148
36	<1	41	26	26	55	16	18	23	30	39	52	69	87	97	203
average for all MSHA Districts							35	41	47	55	65	76	88	97	132
standard deviation							17	17	16	14	12	10	6	2	62

Notes: The incombustible content is the value measured by the MSHA Mt. Hope laboratory.
The soluble content is the percentage that is soluble in hydrochloric acid (i.e., the calcium carbonate content of the limestone rock dust), as measured at OMSHR.
The ash includes the ash in the coal plus the insoluble mineral material, as measured at OMSHR.

Table B-2. Analyses of size of coal dust particles from return airways for seven Pittsburgh seam coal mines

States	Mines	Samples	-270 mesh or < 53 µm, %	-200 mesh or < 75 µm, %	-140 mesh or < 106 µm, %	-100 mesh or < 150 µm, %	-70 mesh or < 212 µm, %	-50 mesh or < 300 µm, %	-40 mesh or < 425 µm, %	D$_{med}$, µm
PA,WV	7	10	56	62 ± 15	67	72	78	85	93	58 ± 39

Appendix C:
Discussion of the Coal Dust and Rock Dust Properties and Experiments

Limestone Rock Dust Inerting Discussion

From 1985 through 2001, numerous LLEM coal dust explosion tests were conducted in the single entry D-drift, and more recently in A-drift (2008), to determine the concentration of rock dust required to prevent explosion propagation as a function of coal dust particle size, volatility, and other related issues (Table C-1 through Table C-3).

During the LLEM tests with the pulverized Pittsburgh seam coal dust (~80% minus 200 mesh or 75 µm), the total incombustible content (TIC) required to prevent an explosion propagation was greater than 79% but less than 81.5%. This determination was based on a series of 12 explosion tests (Table C-4) [Cashdollar et al. 1987; 1992a,c; Weiss et al. 1989; Greninger et al. 1991; Sapko et al. 1989; 1998; 2000]. In two of these tests (LLEM tests #51 and #401), the flame ended well within the dust zone. In the three tests (LLEM tests #70, #255, and #386) where the TIC was 79%, the flame travel extended to or slightly beyond the end of the dust zone. The other 7 tests resulted in flame travel well beyond the dust zone. Non-propagation is defined as no sustained flame propagation of the dust mixture. Propagation is defined as flame propagation of the dust mixture.

During the LLEM tests with the coarse Pittsburgh seam coal dust (~20% minus 200 mesh or 75 µm) [Sapko et al. 1989; Weiss et al. 1989; Greninger et al. 1991], a 70% TIC dust mixture prevented an explosion propagation (LLEM test #191). A TIC of ~68% resulted in a propagating explosion (LLEM test #71).

Prior to having recent access to the MSHA band samples collected from underground coal mines throughout the country, there was growing evidence from limited dust surveys that the coal dust particle size had been decreasing since the promulgation of the existing rock dusting regulations. This decrease occurred as new mining technologies were adopted by the industry (e.g., mining methods involving increased mechanization). For this reason, several explosion tests involving intermediate–sized coal dust particles were conducted within the LLEM. One test (LLEM test #88) involved the use of medium-sized Pittsburgh seam coal dust (~45% minus 200 mesh or 75 µm). To achieve this coal dust blend, pulverized coal dust was added to the coarse dust. For this single test, the medium-sized coal dust was mixed with rock dust to result in a ~67% TIC for the coal/rock dust mixture. Upon ignition of the methane zone, this mixture resulted in a propagating explosion.

Additional tests were later conducted with a blend of pulverized and fine coal dust to provide an average coal dust particle size ranging from 83% to 85% minus 200 mesh or 75µm. This pulverized-fine dust mixture, when mixed with rock dust to result in a ~79% TIC dust mixture, resulted in a propagation (LLEM test #357 and #387). A non-propagation resulted using an 81.6% TIC pulverized-fine coal dust mixture (LLEM test #353). The results from these tests were similar to the tests with the pulverized coal (80% minus 200 mesh or 75µm).

One additional test (LLEM test #388) was conducted with a finer Pittsburgh seam coal dust (95% minus 200 mesh or 75 µm). A propagation resulted after using a ~79% TIC fine coal dust mix.

Based on the LLEM explosion tests, the coal dust particle size has a substantial impact on the propagation potential for a coal dust. As the coal dust particle size decreases, increasing amounts of rock dust are necessary to render the coal/rock dust mixture inert. The greatest impact is evident between the particle size of the coarse (20% minus 200 mesh or 75 µm) coal dust and the pulverized (80% minus 200 mesh or 75 µm) coal dust. To ensure non-propagation within the

LLEM, the coarse coal dust required at least 70% TIC and the pulverized coal dust required greater than 79% and less than 81.5% TIC.

During the first test (LLEM test #517) with the medium-sized coal dust (38% minus 200 mesh or 75 μm), a 74% TIC dust mixture resulted in a propagation. Two tests (LLEM tests #518 and #522) were conducted with a ~76% TIC dust mixture and resulted in a non-propagation. The results of these medium-sized coal dust inerting tests are summarized in Table C-4.

Table C-1. Pittsburgh seam coal dust sizes

Size	Year	−400 mesh or < 38 μm, %	−200 mesh or < 75 μm, %	−100 mesh or < 150 μm, %	−50 mesh or < 300 μm, %	−30 mesh or < 600 μm, %	D_S, μm	D_W, μm	D_{med}, μm
Pittsburgh seam bituminous coal from PA									
Coarse	1980s	9	19	30	43	65	96	440	380
Coarse	2008	10	20	34	53	82	84	320	270
Medium	2008	16	38	61	79	100	55	166	104
Pulverized	1980s	41	80	99	100	100	28	48	45
Pulverized	2008	30	69	98	100	100	33	60	57
Pulverized-fine	1999	64	85	97	100	100	21	40	31
Fine	1999	86	95	98	100	100	17	24	<38

Table C-2. Limestone rock dust sizes

Size	Year	−400 mesh or < 38 μm, %	−200 mesh or < 75 μm, %	−100 mesh or < 150 μm, %	−50 mesh or < 300 μm, %	−30 mesh or < 600 μm, %	D_S, μm	D_W, μm	D_{med}, μm
Pulverized	1980s	62	76	95	100	100	14	47	24
Pulverized	2007	54	72	98	100	100	10	51	26

Table C-3. Average proximate and ultimate analyses of coal used in the LLEM experiments

	Pittsburgh Coal As received, %
Proximate analysis	
Moisture	1.7
Volatile matter	36.5
Fixed carbon	55.6
Ash	6.2
Total	100.0
Ultimate analysis	
Hydrogen	5.4
Carbon	77.4
Nitrogen	1.5
Oxygen	8.1
Sulfur	1.4
Ash	6.2
Total	100.0

Heating value = 13,803 Btu/lb

Table C-4. LLEM inerting tests for Pittsburgh seam coal dust and limestone rock dust using a 40 ft long ignition zone

LLEM test no.--entry	Date	Coal dust					Rock dust, %	Total Incombustible, %	Flame travel, ft	Result
		Size	−200 Mesh, %	Zone, ft	Conc., g/m³					
49–D	7/17/85	pulverized	~80	40–250	200	70.0	72.3	750	P
50–D	7/25/85	pulverized	~80	40–250	200	75.0	77.1	500	P
51–D	8/1/85	pulverized	~80	40–250	200	80.0	81.5	200	NP
53–D	9/4/85	pulverized	~80	40–640	200	75.0	77.1	750	P
69–D	4/24/86	pulverized	~80	40–250	200	73.0	75.2	600	P
70–D	5/1/86	pulverized	~80	40–250	200	77.0	78.8	300	P
71–D	5/8/86	coarse	~20	40–250	200	65.0	67.8	390	P
77–D	8/6/86	coarse	~20	40–250	200	50.0	54.0	500	P
83–D	10/9/86	pulverized	~80	40–250	200	65.0	67.8	750	P
87–D	11/20/86	coarse	~20	40–250	200	60.0	63.2	600	P
88–D	11/25/86	medium	~45	40–250	200	65.0	67.2	750	P
90–D	1/8/87	pulverized	~80	40–430	200	65.0	67.8	750	P
190–D	6/21/89	coarse	~20	40–310	200	73.0	75.0	175	NP
191–D	7/12/89	coarse	~20	40–310	200	67.7	70.0	200	NP
255–D	1/16/91	pulverized	~80	40–490	200	77.2	79.0	445	P
352–D	9/30/97	pulv/fine	~83	40–250	200	83.0	84.4	150	NP
353–D	10/27/97	pulv/fine	~83	40–250	200	80.0	81.6	200	NP
357–D	12/17/97	pulv/fine	~83	40–250	200	77.0	78.8	300	P
386–D	9/8/99	pulverized	72	40–310	200	77.0	78.8	300	P
387–D	9/15/99	pulv/fine	85	40–310	150	77.0	78.8	300	P
388–D	9/23/99	fine	95	40–310	150	77.0	78.8	300	P
398–D	3/1/01	pulverized	~80	40–460	200	65.0	67.2	750	P
401–D	3/28/01	pulverized	~80	40–460	200	80.0	81.6	200	NP
512–A	1/9/08	pulverized	69	40–340	200	75.0	77.0	355	P
513–A	1/15/08	pulverized	69	40–340	200	80.0	81.5	230	NP
514–A	1/23/08	coarse	20	40–340	200	64.0	66.9	355	P
516–A	2/6/08	coarse	20	40–340	200	69.0	71.5	280	NP
517–A	2/13/08	medium	38	40–340	200	71.7	74.0	355	P
518–A	2/27/08	medium	38	40–340	200	74.4	76.4	280	NP
520–A	3/12/08	medium	38	40–340	200	68.5	71.0	550	P
522–A	3/26/08	medium	38	40–340	200	74.4	76.4	280	NP

Effect of Particle Size on Coal Dust Explosibility

The effect of coal dust particle size on explosibility is illustrated in Figure C-1, which contains data collected from large-scale explosions conducted in the LLEM from the 1985 through 2008. This curve shows the amount of incombustible material required to prevent propagation for coal dust containing 20% to 85% particles passing a no. 200 sieve (< 75 μm). Given the experimental test conditions, the curve is the boundary between mixtures that did propagate an explosion (below line) and mixtures that did not propagate an explosion (above line). Experimental results also show that the TIC required to prevent flame propagation becomes much less dependent on coal particle size as the TIC approaches 80%.

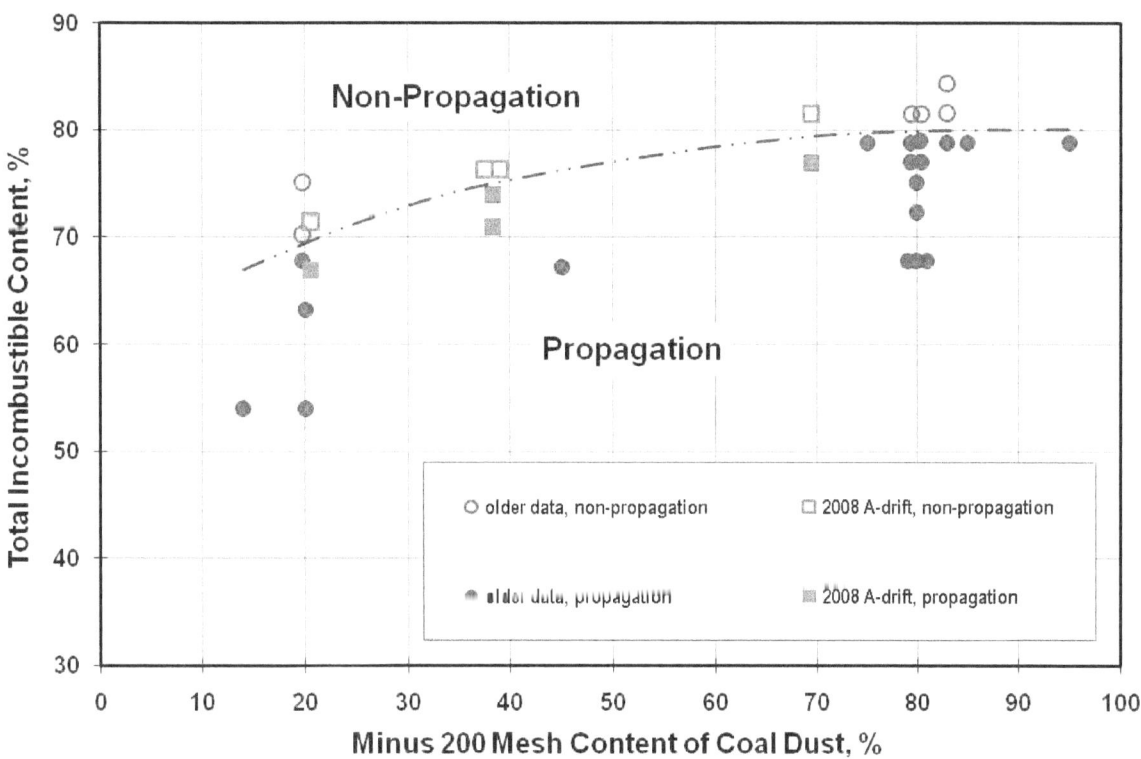

Figure C-1. Effect of particle size of coal dust on the explosibility of Pittsburgh seam bituminous coal as tested within LLEM.

www.ingramcontent.com/pod-product-compliance
Lightning Source LLC
Chambersburg PA
CBHW081902170526
45167CB00007B/3117